The Origin of Kinds,

By Means of Creator God
And
The Preservation of Souls in the Struggle for Eternity

**Dr. Anthony R. Silvestro, Jr.
Jonathan Eckel, Ev.**

Foreword: Pastor Andrew Rappaport

Unless otherwise indicates, all scripture quotations are from the ESV Bible (The Holy Bible, English Standard Version), copyright 2001 by Crossway, a publishing ministry of Good News Publishers. Used by permission. All rights reserved.

Cover created by Matt Yokom

On The Origin of Kinds By Means of Creator God And The Preservation of Souls in the Struggle for Eternity

© January, 2016, Anthony R. Silvestro, Jr. Except as provided by the Copyright Act, no part of this publication may be reproduced, stored in a retrieval system or transmitted in any form or by any means without the prior written permission of the publisher.

Remember Creation Publishing
First Edition, corrected, Soft Cover- 2017

ISBN 978-0-9994962-0-6

Dedication

I am so thankful for my wife, Julie, and son, Anthony III, and their never-ending support. I am also so thankful for everyone that the Lord has placed into my life in my growth in God's Word.

This book would not have been possible without the help from many individuals. A special thanks goes out to those who have taken many hours to help to shape, edit, and proofread this book:

Pastor Andrew Rappaport, Striving For Eternity Ministry
Pastor Chris Hinckley, Olmsted Falls Baptist Church
Pastor Steve DeCoste, Olmsted Falls Baptist Church
Pastor Austin Hetsler, Christ the Rock Church
Scott Weckerly, Creation Speaker

Dr. Anthony R. Silvestro, Jr.

I would like to thank my amazing wife and all my beautiful children. I pray this book teaches them and many others how to understand what we believe as Christians and how to defend biblically what we believe. I pray that they will always stand on the word of God with out apology.

Jonathan Eckel, Ev.

Preface

It has been my objective to write a Biblically sound evangelism book. While many fantastic books exist that teach Biblical evangelism, presuppositional apologetics, and creation apologetics separately, there are not any that put everything together into one easy-to-apply book. In this book, you will learn how to Biblically evangelize, using the Law and the Gospel, and to handle the most common objections using presuppositional and creation apologetics.

The book will also address one of the biggest lies today- we will demonstrate why evolution is wrong and why the Christian evolutionist must attempt to serve "two masters" – man and God. The battle cry of the Reformation was the 5 *Solas:*

Sola Fide = **By faith alone.**
Sola Scriptura = **By Scripture alone.**
Solus Christus = **Through Christ alone.**
Sola Gratia = **By grace alone.**
Soli Deo Gloria = **Glory to God alone.**

The Christian evolutionist may say that these are correct, yet evolution and billions of years are nowhere to be found in Scripture. The problem is that, to be a *consistent* Christian, we must believe the entire Word of God, starting at the beginning of Genesis. This book will demonstrate that we must start with the correct Christian Worldview to make sense of the entirety of Scripture and be *consistent in what we believe.*

Many doctrines in Scripture, starting with the Gospel (1 Corinthians 15:3-4), are undermined when a Christian does not believe what God plainly says in the beginning of Genesis. The problem stems from the fact that the starting point for a Christian evolutionist is not actually Scripture. It does not take too long to figure out that the starting point for them is from *some* of the science of today – not God. In the end, the Christian evolutionist is only serving man.

Dr. Anthony R. Silvestro, Jr.

I pray that as you progress through the pages of this book, you have a true understanding of why this is such a pivotal issue for a Christian to take a stance on. I also want you to know that we have written this book for one purpose. That is simply, to bring Glory to God!

This issue can often be looked down upon because many are saved without understanding the main issue, which is the way you view scripture. If you can trust the scriptures for the truth on how to get saved, how then can you not trust them when they speak of creation? I pray this helps draw you into a deeper understanding on why it is so important to hold to what God says in and through His Word. His infallible and inerrant word is sharper than any two-edged sword.

Jonathan Eckel, Ev.

Why Evangelize?

Have you ever heard this quote? (emphasis added)

> "I've always said that I don't respect people who don't **proselytize**. I don't respect that at all. If you believe that there's a heaven and a hell, and people could be going to hell or not getting eternal life, and you think that it's not really worth telling them this because it would make it socially awkward—and atheists who think people shouldn't proselytize and who say just leave me along and keep your religion to yourself—**how much do you have to <u>hate</u> somebody to *not* proselytize? How much do you have to <u>hate</u> somebody to believe everlasting life is possible and not tell them that?** I mean, if I believed, beyond the shadow of a doubt, that a truck was coming at you, and you didn't believe that truck was bearing down on you, there is a certain point where I tackle you. And this is *more* important than that…"

Whether you have or have not heard this quote, which biblical evangelist do you think said it? Here are your choices:

    A. Ray Comfort
    B. CH Spurgeon
    C. Martin Luther
    D. John Calvin
    E. None of the above

If you answered "E", then you would be correct! But then, which other super-evangelist may have said this? None of them. The correct answer:

Atheist, Penn Jillette, from the world famous comedy duo, Penn & Teller.

That is right- *a professing atheist*. Penn uploaded a video with this quote in 2009 after he was handed a Gideon's Bible from an evangelist. As a professing atheist, Penn has an understanding regarding biblical evangelism that so few evangelicals have - **the understanding that every person who dies before repenting and trusting in Jesus alone will go to hell. For eternity.**

That statement should give every evangelical the urgency to get out and share the "Good News." If we truly love people, we should be warning them of the impending lake of fire – not just walking past them while exchanging "niceties." All too often, professing Christians criticize biblical evangelism because they claim that we are not "showing people the 'love of God'." All we are doing by preaching the law, showing them their sin, and calling them to repent and believe the Gospel is upsetting people. We need to remember to direct them to what Jesus said in John's Gospel:

"For God so loved the world, that he gave his only Son, that whoever believes in him should not perish but have eternal life." (John 3:16)

God manifested His love for us in *the death of his Son on the cross*. He took our sin so that we could have eternal life.

Purpose

The purpose of this book is to teach how a Christian is to evangelize *biblically* and to be able to answer the most common objections to the Christian faith - especially those regarding origins and Jesus being the only way. **This only happens when we start with the Christian Worldview.**

It is not just the "Christian Apologist" that is responsible for defending the Christian worldview – every believer has the same obligation. Peter makes this abundantly clear:

"Finally, **all of you,** have unity of mind, sympathy, brotherly love, a tender heart, and a humble mind… Now who is there to harm you if you are zealous for what is good? But even if you should suffer for righteousness' sake, you will be blessed. Have no fear of them, nor be troubled, but in your hearts honor Christ the Lord as holy, **always being prepared to make a defense to anyone who asks you** for a reason for the hope that is in you; yet do it with gentleness and respect." (1 Peter 3:8, 13-15)

Every believer is given the charge to be an "apologist." We are called to provide answers to the honest questions that we are asked about the truth of Christianity. **And every believer is responsible for sharing the Gospel.**

Table of Contents

Forward: (Pastor Andrew Rappaport)

Introduction: (Jonathan Eckel)

Chapter 1:
The Christian Worldview (Dr. Anthony R. Silvestro, Jr.)

Chapter 2:
Origins (Dr. Anthony R. Silvestro, Jr.)

Chapter 3:
Basic Evolutionary "Science" (Dr. Anthony R. Silvestro, Jr.)

Chapter 4:
The Relevance of Genesis (Dr. Anthony R. Silvestro, Jr.)

Chapter 5:
Creation, the Fall, and the Promise (Dr. Anthony R. Silvestro, Jr.)

Chapter 6:
The Gospel (Jonathan Eckel)

Chapter 7:
An Introduction to Presuppositional Apologetics (Dr. Anthony R. Silvestro, Jr.)

Chapter 8:
Problems with Evolution (Dr. Anthony R. Silvestro, Jr.)

Chapter 9:
The Reliability of the Bible (Dr. Anthony R. Silvestro, Jr.)

Chapter 10:
Textual Variants (Pastor Andrew Rappaport)

Chapter 11:
How the Gospel is Affected by the Evolutionary Argument (John Eckel)

Chapter 12:
A Call to Repentance (Jonathan Eckel)

Chapter 13:
Basic Challenges – Age of the Earth (Dr. Anthony R. Silvestro, Jr.)

Chapter 14:
Basic Challenges – Part 1 (Dr. Anthony R. Silvestro, Jr.)

Chapter 15:
Basic Challenges – Part 2 (Dr. Anthony R. Silvestro, Jr.)

Chapter 16:
Strategy of an Encounter - and How to Use This Knowledge (Dr. Anthony R. Silvestro, Jr.)

Quick Quips

Footnotes

Forward

There are many books on the market that discuss evangelism, creation science and presuppositional apologetics, but there are few that tie these three topics together. Many of the books on the market leave the reader with the thinking that these are three separate and distinct topics that either have nothing to do with each other, or, even worse, the thinking that they are mutually exclusive. *On the Origin of Kinds* is unique that it brings together these three topics together to provide the reader with an apologetic approach to sharing the gospel and being ready to answer questions for the faith that lies within us.

We live in a culture that has become hostile to the things of God. Much of the world benefits from the Christian values that influenced its culture, yet the majority of people now display their hatred towards God. This is seen in their desire to change anything and everything that reflects Biblical principals, even if it leads their culture to destruction. We see a global attempt to destroy or redefine marriage, to confuse over basic things like gender, and redefine tolerance to mean totalitarianism.

We have moved from a culture where people would pretend to be Christian to fit in to one where Christians attempt to fit in by pretending to be like the world. The church has stopped trying to be the lighthouse to a lost and dying world shining the light of the gospel into one that tries to be culturally relevant with 30-year-old references in sermons and programs. Pastors think they are being hip and relevant by having sermon series based on 1980's or 1990's movies and TV series. It is no wonder why so many Christians question why the modern church is so anemic! The church has little to no impact influencing the culture because they are trying to be relevant to the culture to reach people using mass marketing as if people were purchasing the church like a product.

Many people have a fear to share their faith because they are scared of that question that they just will not have an answer to. In this work, it provides not only the basics of sharing the gospel but has much for the more advanced evangelists. This book is one of the few that you will want to read multiple times to obtain all the material at your different stages of maturity. Charles Darwin's book influenced generations of unbelievers with falsehoods in their hatred for God. This book has the potential to influence generations of believers with the truth in their love for God and their neighbor.

Dr. Anthony Silvestro has spent many years in the past arguing against Christianity based on the secular scientific acceptance of information. When he became a Christian, he did not give up on his background in science. In fact, he used his background to inform his Christianity, but he had not given up on his long held belief in evolution. Dr. Silvestro tried reconciling the Bible and evolution until he was challenged to question his religious doctrine of scientism (using science as a religion). Being faced with all of the science, he realized that the Bible was accurate from beginning to end! Good science existed for a great number of subjects contained in the Bible, starting with the creation of the universe and everything in it!

Johnathan Eckel has been on the streets sharing the gospel for a long time. He has heard all the arguments, but he spends his chapters of this book to get to the heart of the issue of the gospel. There is no fluff; he gets to the important issues that must be understood to rightly communicate the gospel to a lost and dying world.

I pray that this book is a blessing to you!

Introduction: Lights, Camera, Action?

Dear Elders,

What makes a church grow? Do you believe that it rests on you and your ability? Is it your skills that successfully reach a crowd? Do you have the expertise to read a crowd and tailor your speech while you speak? Are you the power behind your church in gaining new members? Maybe it is the amazing Sunday School that your church provides, or the free "worship concert" provided every week? Sorry to break the news to you- only Christ builds the church. And the Bible gives us the means on how to do it.

So how does a church grow *biblically*? As we take a look at this "popular" subject, we have to remember that, as Christians, we must always start and stand on God's Word -The Bible- for all of our views (chapter 1 addresses this). This is no different, although you might get confused while walking around a "Christian" bookstore. You will see all kinds of "strategies" and different "ways" to *manipulate, trick, bribe or entertain* people into coming through your doors. One of the things that you have to remember is that **it is not your doors**. The doors (as well as the rest of the church) belong to Christ. He gave you the lamp stand. If He chooses to grow the body of believers in your church, then that is His prerogative to do so.

It has often been said, "What you draw them with, is what you keep them with." Far too many churches today try to use gimmicks and different types of music styles to make it seem like you are going to a concert put on by the world. Think about it - how many times have you heard a variation of one or more of the following statements:

"Come to my church because the music is awesome!"

"Come to my church because my kids love the children's programs!"

"Come to my church because the pastor is funny!"

"Come to my church because the worship music is like a cool concert!"

"Come to my church because the café is great to hang out in!"

Yet, nowhere in the Bible will you see any of these, nor any other, *worldly* ways, being used to grow a church. Nor will you find the insinuation that is given in some books: "Make sure you buy a fog machine and a powerful sound system to bring in potential new members."

There is, however, a biblical way to plant, run, and grow a church. God has already given us the instructions on how to do this - in the book of Acts. When you read through this book, you will see many points of interest on this subject. Here are just five that I am going to highlight, in no particular order. (This is not an exhaustive study, so I would suggest that you do a more thorough one on your own.)

1. We are to fellowship with those that are in our church.
2. We are to pray with each other, for each other, and for the church.
3. We are to be under strong, unapologetic, expository preaching.
4. We are to be His (Jesus') witness among the people through *Biblical* evangelism. **Through this evangelism, believers will be added to the body**
5. The discipling of new believers by older ones, which fulfills the Great Commission

The first in the list is *fellowshipping* with those in our church. We should have contact with those in our local body so that we can bear each other's burdens and be praying for each other. In too many churches these days, many people keep to themselves thinking that they can either handle things all on their own, or they don't want to be transparent regarding their lives. Sadly, in almost every case, that is a result of pride - they need to repent because they are sinning. Some do it because they are nervous to open up to others, but they also need to repent because they are being disobedient. We are told to mourn with those who mourn and weep with those who weep. Upon further consideration, there is yet another possibility, they are living in outright sin and don't want it exposed. Willful unrepentant sin, according to 1 John, means that the person is of the devil and not saved. Fellowship is vitally important and far too often overlooked in today's modern church.

The second point is in regards to prayer. How serious do you take your prayer life? Do you only do it when you can fit it into your busy schedule? Do you use the common excuse, "I just can't seem to find time!" That is foolish! Many find the time to watch television. Did you know that children aged 2-11 watch over 24 hours of TV per week, while adults aged 35-49 watch more than 33 hours in that same time period? This has become the modern family god (idol). They devote their time and selves to it.

Praying for each other is an important part of fellowship as well. It shows that you care, and we are told by our Lord to pray. Study the life of Jesus and you will see that prayer earmarks His life. The four Gospels *scream* of constant prayer!

The third point is sitting under strong, unapologetic, expository preaching. Before I start, I have to say that a topical sermon from time to time is not bad. But that should NOT be the regular sermon at church. The preaching pastor should be proclaiming and teaching the Word of God every week. Expository preaching keeps God's Word as pure as possible and *taken in context*. This is best accomplished when teaching through God's Word verse-by-verse. This expository preaching communicates God's thoughts to us as unadulterated as possible. Topically based preaching tends to cherry-pick verses out-of-context to conform to the thoughts of the preaching pastor.

God's Word is to be used to help us be conformed to the image of Christ more and more throughout life. God's word is sharper than any two-edged sword. It cuts deep and does not return void - He uses this for His purpose. Not to mention, the *Crescendo* of the Bible -The Gospel - should be preached every week. There has never been a time I have heard the Gospel and thought, "This is getting old".

When the first three points of the list are put into practice, it breeds the environment for points 4 and 5. (Good ortho*doxy should* result in good ortho*praxy*). Effective biblical teaching should impart a love in our hearts for the lost. **Remember, those who have not repented and trusted in Christ alone before they die go directly to hell- no passing "go" and collecting a second chance.**

Evangelism, as spoken of in point #4, is the mode or vehicle that God uses to grow His church. This is explicitly laid out in the great commission:

"And Jesus came and said to them, 'All authority in heaven and on earth has been given to me. Go therefore and make disciples of all nations, baptizing them in the name of the Father and of the Son and of the Holy Spirit, teaching them to observe all that I have commanded you. And behold, I am with you always, to the end of the age.'" (Matthew 28:18-20)

When we go out to spread the *Glorious* Gospel, we should be praying for those that we would speak with to be saved, if it is God's will. **It is those people whom God saves that are perfect for you to invite to your sound, Biblical church.** It is also a great idea to leave your email with them for any questions that they may have or for prayer requests for them.

When you witness (share the truth of the Gospel) to people, you will encounter a few different types of individuals. If you study Acts 2 and Acts 17, you will see the Gospel presented in a bit of a different way to the two extreme groups of people. Please realize that the Gospel is not changed, but that the witnessing that comes before it is. The reason is simple. The people being witnessed to in Acts 2 by Peter already knew the Law (The Ten Commandments) and the fact that they were sinners. They studied and knew the Scriptures well, and they knew of the prophecy for the coming Savior (Messiah). They were waiting for the Him who would fulfill the prophecy, and Peter's "job" was to show that Jesus is the Messiah that they were waiting for. The ones being witnessed to in Acts 17 by Paul are pagans and believe in many false gods, including having a statue of the unknown "god." Paul could not just start by preaching Christ and Him crucified. They had no knowledge of their sin to understand their need for a Savior!

Instead, Paul realized that he needed to go back to the creation account, speak about sin, and *then preach Christ crucified.* We have to realize that every person we share the Gospel with falls somewhere along this continuum.

Even though the church is not meant for unbelievers, there are many churchgoers that have heard the Gospel over and over again and they are not saved. If some are newer to your church, they could fall anywhere between the people in Acts 2 and Acts 17. That's why you need to continually preach the Gospel every Sunday. Those of the world cannot tell you much, if at all, of the Christian Faith. They do not understand that they have sinned against a Holy, Righteous, and Just God, and they are destined for hell if they do not repent and believe the Gospel.

Remember why we share the Gospel: it is entirely for God's glory that people might be saved, and that the love of Christ should compel us!

> "For the love of Christ controls us, because we have concluded this: that one has died for all, therefore all have died; and he died for all, that those who live might no longer live for themselves but for him who for their sake died and was raised. From now on, therefore, we regard no one according to the flesh. Even though we once regarded Christ according to the flesh, we regard him thus no longer. Therefore, if anyone is in Christ, he is a new creation. The old has passed away; behold, the new has come. All this is from God, who through Christ reconciled us to himself and gave us the ministry of reconciliation; that is, in Christ God was reconciling the world to himself, not counting their trespasses against them, and entrusting to us the message of reconciliation. Therefore, we are ambassadors for Christ, God making his appeal through us. We implore you on behalf of Christ, be reconciled to God. For our sake he made him to be sin who knew no sin, so that in him we might become the righteousness of God." (2 Corinthians 5:14-21)

The fifth and final point that was listed is concerning discipleship. This is vitally important for every person in the church -

especially the new believers. This is crucial to the spiritual growth of those you surround yourself with in the church body. In Titus 2, older men are instructed to help the younger just like the older women are told to help instruct the younger. We are to pour into each other and to go through this fleeting life together. We were made to do this life together. Otherwise, there would be no reason for the Lord to say this in Proverbs:

> "Iron sharpens iron, and one man sharpens another." (Proverbs 27:17)

Community and discipleship are vitally important to a Christian's life. Not only because we are commanded to in the great commission but also because it is good for us. It is necessary to help us grow and hold each other accountable.

This book is meant to help train believers on how to fulfill the great commission. In this book, you will:

1. Learn how to share the Gospel biblically,
2. Learn how to defend your biblical worldview properly while dismantling all other worldviews,
3. Learn how to use presuppositional apologetics and creation apologetics to answer the most commonly asked questions and challenges, and
4. Start to disciple new believers in the Christian worldview.

Again, I ask the question. What makes a church grow? Do you believe that it rests on you and your ability? Is it your skills that successfully reach a crowd? Do you have the expertise to read a crowd and tailor your speech while you speak? Are you the power behind your church in gaining new members? *If so, Repent! Christ builds the church, not you!*

This entire book is dedicated to help you equip the Saints- your sheep- to share the Gospel biblically. Enjoy!

Chapter One

The Christian Worldview

Is This Not Just an Evangelism Book?

This book starts "at the beginning." It is designed to make sure that we all have the same **correct** starting point. As we will see throughout this book, a person's starting point will shape his conclusion about everything that he assesses. Two different people can evaluate the exact same thing, but they will come to different conclusions based on their differing starting points. Look no further than an evolutionist and a creationist who would be evaluating the exact same dinosaur bone, but come up with completely different conclusions.

What is the correct starting point? This book will establish that Christianity is the **only** worldview that can correctly evaluate the world. But not only that - this book will also establish what is the precise and accurate Christian worldview. So much of man's opinion has tainted and/or misrepresented what Christ actually teaches in His Word, the Bible.

Worldview Definition

What is a *worldview*? To answer this question, we must first understand that a worldview is made up of a set of building blocks that are called *presuppositions*. These presuppositions are certain beliefs that are assumed from the beginning (things that we *pre-suppose* to be true). We will use a few illustrations to help with the understanding of a worldview.

Most people have heard the phrase, "looking at the world through rose-colored glasses." It means that the person "wearing" these rose-colored glasses looks at the world with more optimism than the average person. Let us say that you are walking through a major downtown city, side-by-side with that person. You may each observe the exact same things as you walk, but your thoughts, reactions, responses, and emotions that result from what you witness *are different* because of the more optimistic nature of that person.

Now imagine that you woke up one Monday morning and decided to go about your day at work and home with a pair of red-lensed glasses on. Everything that you saw would have been in a shade of red. Now, on Tuesday morning, you decide to go through the exact same routine as Monday, but with blue-lens glasses on. Everything you see, the same stuff you saw on Monday, is now shaded blue. You viewed the same objects on both days, but they appeared *different.* Your brain interpreted them differently because of one thing that changed- the color of the lenses in your glasses.

Taking this a step further, imagine that your worldview is a pair of eyeglasses that you never take off, and the presuppositions are the components of the eyeglasses- the frames and the lenses. *Everything that you observe about the world is through those eyeglasses.* With every observation you make through those lenses and the frames that hold them, you have some assumptions that you must assume to be true from the beginning. Some of your starting beliefs about the glasses that you wear are that the lenses are completely clear, they are the correct prescription, and the frames are holding the lenses at the correct distance from your eyes. If your lenses were actually orange, or the prescription was incorrect, then you could not view the world accurately. **You must be biased and assume that your glasses are correct *from the beginning,* in order to make sense of anything that you observe.**

A person's worldview is composed of numerous presuppositions. These presuppositions are a result of many factors, including: education, parental upbringing, culture, nationality, past experiences, and gender. **Every person has a worldview (a set of glasses) made up of all of his presuppositions, and he uses it to interpret and understand the world around him**. Two people can observe the *exact same thing*, but they will never be in complete agreement about the interpretation of it due to the differences in worldview.

The reality is that not one person does anything, like making an observation or entering a conversation, from a completely *neutral* position. The act of "claiming neutrality" immediately makes that person biased to a presupposition. Claiming "neutrality" is not a neutral position and is, therefore, self-refuting. "Neutrality" is not an option, as we will discuss in Chapter 7.

To better understand these concepts, we are going to examine the Christian worldview.

The Christian Worldview

What is the set of presuppositions that make up the Christian worldview? Quoted from Pastor Andrew Rappaport, "God exists, and He has spoken." Thus, it could be stated that these fundamental beliefs, the starting points of a Christian, are:

1. God exists,
2. The Bible is His Word[1], and therefore, completely true[1].

Every born-again Christian should interpret the world around him using these as his *starting points*. Any professing Christian who does not start with these pre-beliefs is devoid of the true Christian worldview, but is possibly unaware of the seriousness of this issue. Why is that? *Because God's Word says so!* With these presuppositions making up the Christian worldview, there are several truths that we can establish:

1. All Scripture is given to us by God

"All scripture is breathed out by God and profitable for teaching, for reproof, for correction, and for training in righteousness." (2 Timothy 3:16)

This verse informs us that **all** Scripture is given (inspired, breathed-out) by God. We can be assured that God used many authors to write down exactly what He wanted, but each author wrote in his own style and personality. Thus, we see differences in the writing styles of authors such as Luke (physician), Peter (fisherman), and David (shepherd and King). This is explained in detail in chapter nine.

Furthermore, we can be assured that this verse applies to both the Old and New Testaments- not just the OT. As summarized by ApologeticsPress.org[2],

"…It seems certain, considering all of the above information: (1) that Paul had earlier quoted Luke 10:7 as Scripture; (2) that Peter referred to Paul's writings as 'Scripture;' (3) that Paul indicated prior to his writing of 2 Timothy that he wrote 'by the word of the Lord' (1 Thessalonians 4:15; cf. Galatians 1:12); and (4) that much of the New Testament already had been written. Thus, 2 Timothy 3:16-17 'can be interpreted as covering the NT as well as the Old' (Ward, 1974, p. 200)."

Matt Slick also addresses this on his website, CARM.org[3]:

"We see that the Old Testament is repeatedly spoken of as being inspired via the numerous references cited above but what about the New Testament? Are the New Testament books inspired as well? The Christian church has always considered the New Testament documents to be inspired. Though in the early church there were some debates on which New Testament books to include in the Bible, God worked through the Christian church to recognize those inspired works. Therefore we now have 27 inspired books for the New Testament. In 1 Cor. 14:37 Paul said, "If anyone thinks he is a prophet or spiritual, let him recognize that the things which I write to you are the Lord's commandment." In 2 Pet. 3:16 Peter said, "as also in all [Paul's] letters, speaking in them of these things, in which are some things hard to understand, which the untaught and unstable distort, as they do also the rest of the Scriptures, to their own destruction." Also, Jesus said in John 14:26, "But the Helper, the Holy Spirit, whom the Father will send in My name, He will teach you all things, and bring to your remembrance all that I said to you." This means that the Lord has commissioned the apostles to accurately record what Jesus had said because the Holy Spirit would be working in them. So, we can see that Jesus promised direction from the Holy Spirit, that Paul considered what he wrote to be the commands of God, and that Peter recognized Paul's writings as Scripture. In addition, since the Christian Church recognizes the 27 books of the New Testament are inspired and since we see internal claims of inspiration in the New Testament, we conclude that inspiration applies to the New Testament documents as well."

2. Jesus is the only way to Heaven

"Jesus said to him, I am the way, and the truth, and the life. No one comes to the Father except through me." (John 14:6)

Jesus says so much in this passage, but it all boils down to one thing: He is the only way to Heaven. There is no other path; there is no other way to the Father. He is the source of all truth; He is the source of everlasting life. **Because Jesus is the only way to Heaven, that means that every other supposed way is false.** Buddha cannot get anyone into Heaven. Neither can the Pope. Neither can any other false god/false savior/prophet/person, such as: Joseph Smith, Mary, the myriad of Hindu gods, Allah, Mohammed, The Watchtower Society, and so on. *It is only through Jesus.* We will learn more about how Jesus is the only way in Chapters 6 and 11.

3. The Word is truth

"Sanctify them in the truth; your word is truth." (John 17:17)

Jesus is speaking to His Father in Heaven in this verse. He is asking the Father to sanctify, which means to make holy and set-apart[4], those that the Father had given Him. How? **By the truth!** But what is Truth? (This is the same question asked by Pontius Pilate in John 18:38, right before His crucifixion.) His Word is truth! As we saw previously in 2 Timothy 3:16, the entire Bible is God-breathed! Because we see that His Word is truth, *we can make the conclusion that the entire Bible is His Word and is true!*

4. Jesus is the Word

"In the beginning was the Word, and the Word was with God, and the Word was God... And the Word became flesh and dwelt among us, and we have seen his glory, glory as of the only Son from the Father, full of Grace and truth." (John 1:1,14)

In the Gospel of John, and particularly these two verses, we know that Jesus is: God in the flesh, the Word, and full of truth. This essentially closes any "loose ends" from the previous three passages. Even with these strong truths from Scripture, the Christian worldview goes much deeper than this:

"The fear of the Lord is the beginning of knowledge; but fools despise wisdom and instruction." (Proverbs 1:7)

"The fear of the Lord is the beginning of wisdom, and the knowledge of the Holy One is insight." (Proverbs 9:10)

"The fear of the Lord is instruction in wisdom, and humility comes before honor." (Proverbs 15:33)

What does the word "fear" mean here? R.C. Sproul explains it as having a "sense of awe and respect for the majesty of God," and that, "If we really have a healthy adoration for God, we still should have an element of the knowledge that God can be frightening."[5] Thus, the healthy adoration for and fear of our Lord is the *beginning* of knowledge. **To know anything at all, we must start with God!** If a person doesn't start from the Bible, God considers that foolish! We understand why it is foolish by the previous verses that we covered- The entire Bible is God's Word, and it is all truth. Furthermore, the Lord states in Colossians:

"...Which is Christ; in whom are hidden all the treasures of wisdom and knowledge." (Colossians 2:2b-3)

Why does God tell us this through Paul? He starts telling us in the very next verse!

> "I say this in order that no one may delude you with plausible arguments." (Colossians 2:4)

We are to make sure to look to Christ for all wisdom and knowledge and not be led astray by the words of fallen individuals. We can be taken captive- led the wrong way- if we do not start with Him, as Paul also states:

> "See to it that no one takes you captive through philosophy and empty deceit, according to human tradition, according to the elemental spirits of the world, and not according to Christ." (Colossians 2:8)

As we saw in Proverbs 1:7, having reverence for the Lord is the *starting point* for any knowledge! It is not the other way around! If the Bible were not true, we could not know anything! **The proof for the existence of God is that unless you presuppose His existence, you cannot *account* for your ability to prove anything at all!** Both our ability to trust our senses to evaluate the world accurately and our ability to logically reason are essential to evaluate the world around us. To be able to prove anything, we *must start* with the Creator who gave us those abilities. As Christians, we can account for the ability to trust our senses and the ability to reason correctly. Evolutionists, who believe that we are the result of billions of random chemical reactions, cannot account for these things! We will be dissecting this *presuppositional argumentation* in chapter 7.

The Christian must live his life with these truths. The Christian worldview should be applied in every facet of life of one who is saved: how he treats his family, how he conducts himself at work, how he views science, and many others.

The purpose of this book is to teach how a Christian is to evangelize *biblically* and to be able to answer the most common objections to the Christian faith - especially those regarding origins and Jesus being the only way. **This only happens when we start with the Christian Worldview.** Presupposing God gives us the abilities; His Word gives us the answers.

Chapter Two

Origins

Why is the Knowledge of Origins Important?

Most people enjoy learning about their parents as they age and learning about their ancestry. Why? It is because your origins help define who you are. Heritage, familial traits, and genetic diseases are just a few reasons why people care about their origins. Ask the person who was adopted without knowledge of his birth mother and father- the vast majority wish they knew more about their origins!

Going beyond our familial ancestry, everyone has the same 4 basic questions about life:

> Who am I?
> Where do I come from?
> What is my purpose?
> What happens when I die?

These questions are all dependent on our origins. The Bible answers all of these questions for us, starting with the absolute beginning. This is given to us in the first book of the bible, the book of Genesis, directly from the only eyewitness available- God.

Origins: Age of the Earth

Whether you are aware of it or not, one of the biggest attacks today against Christianity is regarding origins - even among professing Christians. Atheism and Agnosticism have used its doctrine of evolution in a major way to remove God from culture, and it has infested the church. To examine origins as a Christian, one must properly understand and stay true to the Christian worldview, as we just examined in chapter 1.

The debate about origins is unlike most other disputes in the church today. Most discussions, like those of eschatology and the mode of Baptism, have the same starting point: the Bible. One can disagree with a brother in Christ regarding one of these issues and walk away from the conversation disagreeing with his opponent's stance but respecting the source of it. The debate on origins, creation v. evolutionism, *is different*. Why?

While the Bible does not give us the original date of Creation, nor explicitly tell us how many years ago it was, it does give us valuable *historical information*. A straightforward reading of the Bible gives you a total of about 4000 years *of history*- 2000 or so years from Adam to Abraham, and 2000 or so years from Abraham to Jesus. You would arrive at these numbers by the chronological information contained in Scripture, especially those given in the genealogies listed in the Bible, including: Genesis 5, Genesis 11:10-32, 1 Chronicles, Jude 1:14, Matthew 1, and Luke 3:23-38.

Knowing that Jesus was born about 2000 years ago using historical records, we can establish that the earth is about 6000 years old. This *young earth* conclusion – a 6000 year-old earth- would be derived using **Scripture as the starting point.**

This debate on origins is different because of the answer to the following question. *What, then, is the starting point if one believes in an old earth?* **It is certainly not Scripture** – instead, it is derived from man's opinion, cloaked in the name of science.

In understanding the genealogies above, the Flood during Noah's day would have occurred around 2300 B.C.[1] It is interesting to point out that there is no ancient culture, according to modern archaeologists, that has so-called verifiable artifacts earlier than an *estimate* of 3000 B.C. Even the Jewish calendar, the oldest known calendar, has the date of creation as 3761 BCE, which would make the earth about 5775 years old. The oldest written artifact is considered to be a Sumerian cuneiform tablet dated in that range[2]. While biblical creationists would disagree with the exact date of the cuneiform tablet, it certainly is in close alignment with a biblical timeline. It should be no surprise that while some scientists believe that humans evolved about 2 million years ago[3], the earliest artifacts would only be in existence for under about 6000 years, but most likely under 4300 years old due to the colossal destruction of the world during the flood, unless it was preserved on Noah's Ark as understood in the flood's historical account in the Bible (Genesis 6-9).

Origins: The Start of all Living Things

The debate regarding the start of all living things, especially the human race, is similar to that of the age of the earth. Using the Bible as our starting point, you now want to know how all the living things started on earth. In reading through Genesis 1, the word "kind" appears *ten* times, including the following two examples:

> "And God said, 'Let the earth sprout vegetation, plants yielding seed, and fruit trees bearing fruit in which is their seed, each according to its *kind*, on the earth.' And it was so." (Genesis 1:11)

> "And God said, 'Let the earth bring forth living creatures according to their *kinds* - livestock and creeping things and beasts of the earth according to their kinds.' And it was so." (Genesis 1:24)

You would easily conclude that God brought forth all living things and said that each would reproduce *after its own kind*. There is no indication of living things reproducing and making a different "kind." It should be noted here that "kind" does not equate to "species." There can be many different species that are of the same created "kind"- both tigers and lions are different species, but of the same "kind." A Biblical "kind" more closely equates to the "order" or "family" in the taxonomic system developed by Linneaeus[4] in the 20th century – a system that was arbitrarily developed to try and classify all living things in some type of system according to physical characteristics.

You would also see that God made man in his own image, and not the result of reproduction from a different biblical kind. The Word is clear that God "formed the man of the dust from the ground" (Genesis 2:7, 3:19), and humans are all of their own Biblical kind – humankind (Genesis 3:20). Being the crowning glory of His creation, He gave man dominion over everything on earth, including all of the other Biblical kinds.

> "Then God said, 'Let us make man in our image, after our likeness. And let them have dominion over the fish of the sea and over the birds of the heavens and over the livestock and over all the earth and over every creeping thing that creeps on the earth.' So God created man in his own image, in the image of God he created him; male and female he created them." (Genesis 1:26-27)

> "…You have made him a little lower than the heavenly beings and crowned him with glory and honor. You have given him dominion over the works of your hands; you have put all things under his feet." (Psalm 8: 4-6)

These concepts are strengthened even more when reading Genesis further:

> "Then God said to Noah, 'Go out from the ark, you and your wife, and your sons and your sons' wives with you. Bring out with you every living thing that is with you of all flesh—birds and animals and every creeping thing that creeps on the earth—that they may swarm on the earth, and be fruitful and multiply on the earth.'" (Genesis 8:15-17)

There is not even a hint of biological evolution contained in Scripture. Once again, the starting point of an evolutionist, even a person branded by the oxymoron "Christian evolutionist," is certainly not Scripture – instead, it is derived from man's opinion in the name of science.

Consider one last point: If the Bible allowed for evolution, then there would have been no point for God to preserve *two* of every kind of animal on the Ark. He could have just evolved every kind of animal again when everyone exited after the flood!

Origins: The Big Bang, or In the Beginning, God Created

The Bible contains the sole historical record of the origins of the universe. Written down by Moses, the only observable evidence of the origins of the universe and the earth was witnessed first-hand by the Creator:

"In the beginning, God created the heaven and the earth." (Genesis 1:1)

The first verse of the Bible says that:

1. God created time.
2. God existed before time (and, therefore, transcends it).
3. God created the heavens.
4. God created the earth.

Reading through the rest of Genesis 1, we see that God created things sequentially. There is a certain order to creation. According to the "big bang" theory, there is a different order for the same created things. These differences are displayed in the following chart:

Genesis	Big Bang / Evolution
Earth before the sun	Sun before the earth
Sea before dry land	Dry land before sea
Sea before atmosphere	Atmosphere before sea
Light on earth before sun	Sun before light on earth
Land plants before sea creatures	Sea creatures before land plants
Birds before dinosaurs	Dinosaurs before birds
Man before death	Death before man

As seen in the above chart, there are significant differences in the order of creation between the "big bang" theory and what we read in the historical account in Genesis. For instance, Genesis records the fact that birds were created on day 4 of creation week, clearly before the dinosaurs were created on day 5. Yet, evolutionists believe that dinosaurs were present on earth for millions of years before they died off and evolved into birds tens of millions of years later.

Just by this one example, it is clear that the order of events in the creation account cannot be reconciled to the order of events for the big bang. One or the other must be true. Furthermore, this presents an insurmountable problem for the Christian who tries to reconcile the big bang and evolution with the Bible. *The sad thing is, rather than believing what God says in Scripture, the compromising Christian would rather believe his own mind.*

The Correct Starting Point

The question regarding origins is this: What is the correct starting point in order to determine the age of the earth?

Conversations with "old earth" creationists (OECs), professing Christians who believe that the earth is about 4.5 billion years old, almost always go the same way. Consider the following actual conversation that recently took place between an OEC and me, which is reminiscent of most of those dialogues.

> Biblical Creationist: The Bible is clear that this earth is only about 6000 years old.
>
> Old Earth Creationist: I used to believe that- I don't anymore.
>
> BC: Why is that?
>
> OEC: I believe that the Bible is not clear on how old the earth is.

BC: Do you believe that the Bible is the inerrant Word of God?

OEC: Yes.

BC: Do you believe that the Bible is the starting point for all knowledge?

OEC: Yes.

BC: More importantly, is the Bible *your* starting point?

OEC: Yes.

BC: What about the genealogies and historical accounts given throughout Genesis and the rest of the Bible? Do you believe that they are correct?

OEC: I don't think that the genealogies are complete...

(At this point, I directed the conversation into some apologetics regarding the age of the earth and a literal 6-day creation, which will be covered later in this book. I then shifted the conversation with the following question.)

BC: So, you believe that the Bible is the inerrant Word of God, but you don't take Genesis literally, you think that the hermeneutic of YECs is wrong, and you believe that the Hebrew is translated incorrectly, etc. All of this leads to the fact that you reject that the earth is only about 6000 years old.

So, how old do you think that the planet earth is?

OEC: About 4.5 billion years old.

BC: The Bible doesn't give any indication that the earth is millions to billions of years old. Where did you get that "4.5 billion" number?

OEC: I agree with what (some of the) scientists say. You know, with radiometric dating methods and stuff...

BC: So, the Bible isn't actually your starting point on this issue - science is.

This conversation shows the exact problem with many issues in the world today - the starting point of so many professing Christians is NOT the Bible. When Christians have to deal with issues like homosexuality, transgenderism, and abortion, they must start with Scripture to do it correctly. Similarly, when a Christian deals with the origins debate, they must start with Scripture. By the time you are done reading this book, it will be clear that a 6000 or so year-old earth is the result of starting with Scripture, if you do not already believe that. It will be equally clear, if not already, that **the starting point of an "old earth" creationist is certainly not from Scripture - instead, it is derived from man's opinion, veiled in the name of science**.

This chapter started with the acknowledgement that *God is the only eyewitness* to the origins of everything in the universe. Yet, we will be seeing in the coming chapters how some scientists have tried their best to corrupt what God has already revealed to us clearly. It reminds us of God's sarcastic, but truthful, scolding of Job:

> "Who is this that darkens counsel by words without knowledge? Dress for action like a man; I will question you, and you make it known to me. 'Where were you when I laid the foundation of the earth?'" (Job 38: 2-4)

Chapter Three

Basic Evolutionary "Science"

I thought Christians only believe in fairy tales...not science!

What is Science?

To understand basic evolutionary science, it is prudent to understand what science is. According to Webster's dictionary, science is defined as[1]:

1. A branch of knowledge or study dealing with a body of facts or truths systematically arranged and showing the operation of general laws.

2. The systematic knowledge of the physical or material world gained through observation and experimentation.

According to The Science Council, science is defined as[2]:

1. Science is the pursuit and application of knowledge and understanding of the natural and social world following a systematic methodology based on evidence.

Now, let us take a look at the definition of science from a Christian school, Bob Jones University[3]:

1. Science is the systematic study of nature, based on observations. If a phenomenon cannot be observed by man's senses, either directly or with the aid of instruments, it cannot

be dealt with scientifically. For example, all speculations concerning ultimate origins are excluded from the lawful domain of science, since the necessary observations cannot be performed. Closely akin to science is technology, the practical application of scientifically acquired knowledge. Science and technology are often intertwined to the extent that it is impossible to teach the one without touching upon the other.

2. True science, diligently studied and carefully taught, will dispel a multitude of myths. Unfortunately, it has become fashionable in today's media to erase the line of demarcation between true science and science fiction. The Christian teacher of science can render a valuable service to his students by reestablishing that line and stressing its significance.

It is interesting to note that all 3 definitions listed here, 2 from secular sources and 1 from a Christian foundation, are very similar in the sense they are all based on a systematic study on observations. It should also be apparent by the definitions that the Christian explanation is much more specific. I pray that you will see why this is important throughout this book. Christian scientists seem to be more specific and strict than most secular scientists, especially in separating good (experimental) observational science from historical (origins) science[4, 5, 6]. While a full discussion is outside of the bounds of this book, the following explanation should make this evident.

2 Types of Science

A proper understanding of science requires us to distinguish between the two types of science that exist:

1. Historical (Forensic) Science, and
2. Experimental (Operational, Observational) Science

What we often find when we read science textbooks or watch scientific TV shows is that the authors constantly switch between the two types of science without making any distinction. This can be very misleading when drawing conclusions from the science presented. It is actually a logical fallacy, called the fallacy of equivocation, when two different definitions of a word are used interchangeably within the same argument. This is what frequently happens with the word science when speaking about origins.

In using **historical science**, the person is trying to figure out something that happened in the past. Because there is no way to actually test the observation- *since it happened in the past*- he must try to use current facts to piece together what had occurred. The best analogy to this is a murder trial where there is no reliable eyewitness account. The prosecution will use evidences to try and draw a conclusion that the defendant committed the murder using this circumstantial evidence. In the process, he is trying to convince the judge and/or jury to convict the suspect based on evidence that is beyond a reasonable doubt. Remember that this is accepted evidence, but not proven. However, the defendant's attorney will use some of the same available evidences, with different interpretations, to try and prove the exact opposite! It is easily shown how the underlying **worldview**, as we discussed in chapter 1, can wholly affect the conclusion of the evidences (prosecution wanting to prove guilt and defense wanting to prove innocence). It should be noted that a person could not prove something that happened in the past with 100% certainty. Think of how many "airtight" murder convictions, *from the past*, have been overturned 15 years later after DNA testing was used on those previously closed, and now re-opened, cases. In each of those instances, it became clear that the evidences used to gain a conviction were interpreted incorrectly by the prosecution and the judge/jury.

In using **experimental science**, the person is trying to test an observation that is repeatable. A few simple examples include: testing an observation that objects fall at the same rate in a vacuum (gravity) or testing that water boils at 100^0 Celsius. **An important concept with experimental science is that it can be used for determining some evidences used in historical science, but these cannot be used to prove something that happened in the past, as mentioned in the previous paragraph.** In order to try and prove things using experimental science, one must employ the scientific method.

The Scientific Method

When we read any definition regarding science and how scientific knowledge is gained, it is always by utilizing the scientific method[7,8]. The secular scientists agree with the Christian scientists on the basic premise of the scientific method, as we can see in articles from different creation magazines, including a thorough one from Creation Ministries International (CMI)[9]. While there are many ways to state the scientific method, the basic outline is as follows:

1. **Observation-** One must make an observation of something
2. **Hypothesis / Prediction** – One must then construct a hypothesis about the observation
3. **Testable** – One must test the hypothesis to see if the observation is true
4. **Conclusion** – One must make a conclusion that the test confirms the original hypothesis
5. **Repeatable** – If the first test was successful, the test must be repeated multiple times to see if the results still lead to the same conclusion
6. **Theory** – Once the repeat tests continue to confirm the original hypothesis, the hypothesis becomes a theory.

Science accepts that a theory resulting from the proper application of the scientific method is *true*, but it can still be proven wrong in the future. Because of this, science does change over time. Therefore, science is constantly testing to prove the negative. There are a myriad of examples that demonstrate this: what foods are healthy to eat, relationship of the sun to the earth, the idea of global cooling (then global warming, then climate change)[10], and the human contribution to the CO_2 in the atmosphere, the ethos.

Now, I must admit that, as a Bible-believing Christian, I love science. It is a gift from God for us to comprehend His creation better. We must understand something, however. **The reality is that science**

always has the potential to change, unlike the Word of God. When the Bible (the wisdom of God) and science (the wisdom of man) conflict, a *consistent* Christian **must believe the Bible.** After all, what does the Bible say about the wisdom of men?

> "And I, when I came to you, brothers, did not come proclaiming to you the testimony of God with lofty speech or wisdom. For I decided to know nothing among you except Jesus Christ and him crucified. And I was with you in weakness and in fear and much trembling, and my speech and my message were not in plausible words of wisdom, but in demonstration of the Spirit and of power, so that your faith might not rest in the wisdom of men but in the power of God. (1 Corinthians 2:1-5)

Evolutionary Theory: Age of the universe

Evolutionists give an approximate age of 14 billion years for the universe[11] and about 4.5 billion years for the earth[12]. However, even these numbers have been constantly changing over time. Over the last 150 years, because of the many complications in the hypothesis of evolution, these numbers have been constantly increasing. Magically, the age of the earth has increased from about 100 million years old in 1862 to the 4.5 billion years of age today.[13] The supposed age of the earth radically increased in the first-half of the 20th century[13], concurrent to evolution being propagated fervently. This can be demonstrated throughout scientific literature. These large figures for the ages of the earth and the universe are calculated using several methods:

1. Checking the age of the oldest objects in the universe through radiometric dating methods,
2. Determining the expansion rate of the universe and tracing it backwards in time, and
3. Using measurements of the cosmic microwave background (CMB) to determine initial conditions of the universe

It should be noted that there are major assumptions being made in each of these methods, and they all have catastrophic problems. On top of that, new discoveries continue to shatter secular scientists previously held beliefs regarding age of the earth and the timeline of life on earth- this causes them to constantly rethink things in their evolutionistic worldview. One of these new discoveries happened when paleontologist Mary Schweitzer and her team broke open a T.rex bone and saw soft tissue and red blood cells! In their evolutionary worldview, they believed the bone to be 68 million years old, yet they thought the soft tissue could not be preserved for anywhere near that long. Did this cause Mary and her colleagues to rethink the age of the bone? No – they instead sought to find explanations of how the soft tissue could somehow be preserved for many times longer than ever imagined.[14]

We will be addressing the problems with the most popular of the 3 methods above, radiometric dating, in Chapter 14. In the same chapter, we also will discuss the Horizon Problem, which is a fatal blow to the CMB method mentioned above.

It is also assumed by some scientists that our fossil record is made up of sedimentary layers that were laid down very slowly over millions of years. They conclude this because of the slow sedimentation rate witnessed today. This is also used as "proof" by them that the earth is old. However, they cannot prove that this is what actually caused our fossil record/layers to form because *it happened in the past.* As Christians, we understand that the sedimentation rate was much faster at one point in the past- during the time of the Worldwide Flood! We look at the same fossil record/ sedimentary layers and conclude something completely different – they were solely the result of the Flood and it fits perfectly in the Biblical timeline of a 6000 year-old earth.

Evolutionary Theory: The Start of All Living Things

What is (biological) evolution? The answer to this question is harder to find than it seems. Why is it hard? Because of the constant changes in evolutionism based on research by both secular scientists and creation scientists, there seems to be no universally accepted definition, other than "change." Even scientific journals and websites, such as the National Center for Science Education, attempt to dance around the definition.[15]

According to the Merriam-Webster dictionary, it is defined as:

1. A theory that the differences between modern plants and animals are because of changes that happened by a natural process over a very long time.

2. A process of continuous change from a lower, simpler, or worse to a higher, more complex, or better state.

According to Biology.about.com, it is defined as:

1. "Biological evolution is defined as any genetic change in a population that is inherited over several generations. These changes may be small or large, noticeable or not so noticeable. In order for an event to be considered an instance of evolution, changes have to occur on the genetic level of a population and be passed on from one generation to the next..."[16]

The interesting thing about these definitions is in what we actually observe happening in the world around us. As we will be discussing in the rest of this chapter, what we observe in life is exactly the opposite of what we should see. We do not see kinds evolving into more complex kinds over time due to environmental changes. We do not see life going towards better survival overall; in fact, we see life being continually driven to extinction! As stated in a well-known secular science journal, National Geographic, "More than 90 percent of all organisms that have ever lived on Earth are extinct."[17]

Evolutionists wrongly teach that one kind can turn into another kind over a long period of time. It is also "believed" that all life on

Earth shares a common ancestor, which first started about 3.5-3.8 billion years ago.

Some scientists believe in the tree of life (Figure 1, www.pinterest.com), which starts with the assumption that pond scum somehow arranged itself into a single-celled organism. We will call this the bottom of the trunk of the tree. As the single-celled organism evolves over time into more complex organisms, each new organism of significant difference starts its own branch coming off of the trunk of the tree. Each branch eventually develops more branches of more complex organisms. This evolution continues to make more branches from branches. Over billions of years, very complex organisms developed which, today, are at the furthest points from the trunk of the tree.

Humans and monkeys are considered separate branches today, but we share the same branch that we split from millions of years ago. It is fair to point out that not too many years ago, some scientists believed we directly evolved from chimpanzees. This is another example of how "science" continually changes due to new discoveries. Today, it is said that humans share a common ancestor of a monkey, but not a direct descendent of the monkeys of today.

This chart is explained using taxonomy, which is defined broadly as "the science of classification, but more strictly the classification of living and extinct organisms—i.e., biological classification." It uses "the methodology and principles of systematic botany and zoology and sets up arrangements of the kinds of plants and animals in hierarchies of superior and subordinate groups."[18] This means that every organism, whether alive or extinct, is classified and organized into distinct groups with other similar organisms and scientifically named. The classifications starts off being very broad, and then they get more specific the further down the chart that the organism is classified.

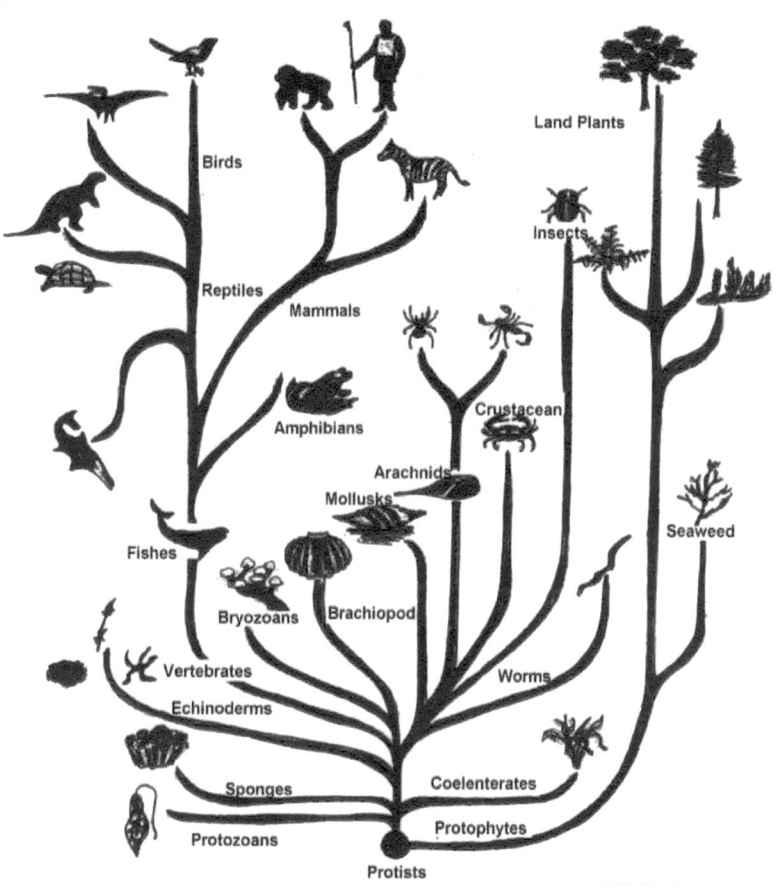

Figure 1

Charles Linnaeus is considered the "father of taxonomy," and is credited with the organizational chart being used today, as seen in the left column of the following table.

	Gray Wolf	Red Wolf	Red Fox	Domestic Dog
Domain	Eukarya	Eukarya	Eukarya	Eukarya
Kingdom	Animalia	Animalia	Animalia	Animalia
Phylum	Chordata	Chordata	Chordata	Chordata
Class	Mammalia	Mammalia	Mammalia	Mammalia
Order	Carnivora	Carnivora	Carnivora	Carnivora
Family	Canidae	Canidae	Canidae	Canidae
Genus	Canis	Canis	Vulpes	Canis
Species	Lupis	Rufus	Vulpes	Familiaris

Sources:

http://www.wolfweb.com/class.html,

https://bioweb.uwlax.edu/bio203/f2013/eidensch_matt/classification.htm

This chart also shows a few examples of how scientists classify organisms- in this case: the gray wolf, the red wolf, the red fox, and the domestic dog. Relatively speaking, the farther down the chart that two organisms are the same, the more evolutionarily related they are thought to be. The problem with this classification system is that there is some *arbitrariness* in how scientists have placed organisms into it in the first place. Similarities among physical features dominate how organisms are classified. In the example above, all four of the species are of the same top 6 classifications. Thus, they are all in the animal kingdom because they are all animals, they are of the class mammalia because they are mammals, and they are of the order carnivora because they all eat other animals. These four animals differ by their genus and species due to some minor variations among them.

As we learned in Chapter 2, God speaks of a different classification system altogether in the Bible. In the beginning of Genesis, it is mentioned that things will reproduce "after its own kind." When we compare the Biblical "kind" with the taxonomic chart above, "kind" is typically equivalent to either the "family" or "order". Using this Biblical definition, the parent "kind" of organism that God both *created in the beginning and preserved on the Ark* contains all of the animals that can interbreed. In the chart above, we can see an example of this- the Biblical canine "kind". We know that all of these animals in the "canine kind", seen above as the family Canidae, can interbreed: the wolf, the fox, and all domestic dogs. Thus, they are all of the same Biblical kind!

Understanding the evolutionist tree of life and the Linneaus classification system above, the type of evolution explained by some scientists is best-termed *general evolution* (evolutionary biologist terminology) or *macroevolution* (creationist terminology). It refers to these major changes over time- the origin of new types of organisms coming from previously existing, but different, ancestral types. In Biblical terms, this would be one of God's created kinds turning into a different kind over a long period of time. An example of this would be dinosaurs changing to birds over tens of millions of years, as is taught in some science classes today. *When someone talks about evolution, they are referring to macroevolution.* This has never been proven, as we will discuss later in this chapter.

A second term to bring up is *special evolution* (evolutionary biologist terminology) or *microevolution* (creationist terminology), not to be confused with *macro*evolution above. Microevolution refers to changes within a given Biblical kind. Change happens within a group, but the descendant is clearly of the same type as the ancestor. An example of this would be using our example above regarding the canine kind; new breeds of dogs have developed over time. The descendent dogs may have some new (better or worse) traits, *but they are still of the canine kind.* We see examples of this throughout the animal and plant kingdoms, where many varieties of apple trees exist (granny smith, red delicious) and various types of felines exist (housecats, panthers, tigers). No scientist disputes special evolution (microevolution), yet it exactly follows what God describes in the Bible.

Evolutionism: The Bait-and-Switch

Secular scientists tell us that there are several observations that bolster evolutionary theory:

1. An extensive fossil record which conforms to the secularist "tree of life"
2. All organisms use the same code for proteins- DNA
3. The many similarities in features among organisms, and
4. Observing "evolution" in action (flu virus changing, peppered moth studies, the Galapagos finches, antibiotic resistance of bacteria)

This last example is what creationists call the "Bait and Switch" – the fallacy of equivocation - by evolutionists. "The fallacy of equivocation occurs when a key term or phrase in an argument is used in an ambiguous way, with one meaning in one portion of the argument and then another meaning in another portion of the argument."[19] In this case, evolutionary scientists only give evidence for microevolution to use as the proof for macroevolution! The fact that evolutionists use a logical fallacy as their only "observable proof" shows how indefensible general evolution is! In their most-used examples:

When the flu virus changes every year, what does it change into? The flu virus!

When the peppered moths changed color, what did they change into? Moths!

When the Galapagos finches were studied in regards to the size of their beaks, what did they change into? Finches!
When bacteria "develop resistance" to antibiotics, what do they turn into? Bacteria!

The examples they give are clearly variations within a kind; there is no change of one kind into another kind! **This is exactly what**

the Bible would predict! We observe changes within a kind, but not between kinds. Again, the dog kind doesn't ever produce a cat kind, or any other kind.

There is simply NO OBSERVABLE EVIDENCE that supports (macro)evolution. This use of the equivocation fallacy is why you will never hear of evolutionists speak using *precise* language regarding these so-called "proofs." But beware: *whenever an evolutionist refers to evolution, they are referring to general evolution (macroevolution) only.* They try to equate the Biblical kind with a species, but it is dishonest on the part of the evolutionist. As we stated earlier, the Biblical kind is most closely associated with the family or order classification.

Evolutionism: Mutations

If you ask an evolutionist how evolution works, you will have any number of the following terms thrown at you, including: natural selection, adaptation, survival of the fittest, variability, hereditability, and genetic drift. All of these concepts boil down to, and are dependent on, one mechanism: genetic mutations. Genetic mutations are real, and scientists acknowledge that mutations affect the organism in one of three ways: a beneficial effect on the organism, a negative effect on the organism, or a neutral effect on the organism.

Of all the **observed mutations** in science, the vast majority of mutations either has no observed effect on the organism (neutral) or has a negative effect on the organism. According to several current published genetic journals, there are more than 6,000 known human single-gene disorders, and they occur in about one out of every two hundred births![19] There are some more staggering statistics regarding the approximately 4 million babies being born each year: about 3 to 4% will be born with a genetic disease or major birth defect, and almost 1% of all babies will be born with chromosomal abnormality, which can cause physical problems and mental retardation.[20] It is easily understood that neither of these types of mutations, neutral nor negative, can produce molecules-to-man evolution. Thus, what is often used as evidence for general (macro) evolution is "beneficial mutations." So what do we actually observe with beneficial mutations?

There are a limited number of examples that we see for beneficial mutations. One is involving the CCR5 gene in humans. A mutation in this gene makes humans virtually immune to HIV. Another is concerning an example used above, the antibiotic resistant bacteria. What is interesting here is that most of these beneficial mutations only cause the organism to have a benefit in a restricted environment. If the organism with the beneficial mutation leaves the restricted environment where it has a perceived benefit, the organism is usually less fit when it is back in the normal environment.

An example of this is concerning sickle-cell anemia. This disease is characterized by a loss of full function of hemoglobin, the protein responsible for transporting life-dependent oxygen throughout the human body. When a person in the United States has this mutation, he is at a genetic disadvantage compared to a "normal" person. He will tend to tire quickly, among other things, due to a restriction of oxygen flow in the body. However, in a restricted environment where deadly malaria is present, this mutation has a beneficial effect. People with sickle-cell anemia are immune to malaria! But, once that person is removed from a malaria-infested environment, he is at a genetic disadvantage.

However, being beneficial is not the only thing that is necessary for mutations to cause molecules-to-man evolution. For general evolution to work, mutations must not only be beneficial, **but they must also *increase* functional genetic information**. Evolutionary scientists believe that atoms magically arranged themselves into a living cell billions of years ago. Then, they believe that mutations that increase functional genetic information piled up over millions of years to make more and more complex organisms. Eventually, enough of these information-increasing mutations gave rise to more complex organisms that resulted in humans. This is displayed in the evolutionary tree of life (figure 1) from earlier in this chapter: the higher up the tree that you go, the more complex the organism is.

Thus, we have now established that once we sift through all of the elusive evolutionary terminology, molecules-to-man evolution relies on one premise: that mutations can increase genetic functional information over time. *Do we actually observe this in science?*

NO!

Of all the mutations that have ever been observed, NONE have ever been shown to increase *functional genetic information*! Every mutation, whether beneficial, neutral, or negative, has displayed a net *loss* of information!

Yet, for general (macro) evolution to occur, you would need billions of mutations that increase functional genetic information over time. This is the main reason why general evolution needs millions-to-billions of years to try and make it work.

One other thing needs to be stated here. Natural selection seems to be the buzzword used today as the mechanism for evolution. It needs to be understood that **natural selection cannot produce new genetic information!** Natural selection only "selects" from the genes that are already present! (Keep in mind that "natural selection" is only a concept. There is no intelligence that makes up natural selection to do the actual selecting.)

So where did all the species come from?

We have established that God created the original kinds in the garden, and we know that He preserved them on the Ark when He destroyed the entire world by water. Since there were only two of each kind (and seven of some) saved by the Ark, then how do we explain where all the different species we see today have come from? All the species that exist today, and those that have existed but went extinct since the floodwaters receded, are real! Remember, they were not on the Ark- only kinds were.

Evolutionists would have you believe that through special evolution (microevolution), new characteristics randomly arise by mutations that add information to the genome. When enough of these mutations, or little changes, get added together, new animals eventually arise (general evolution, macroevolution). They want you to believe that this is what has lead to all the speciation we see today. We must point out again that there has never been a natural process discovered by which new information has been added to the genome! **Since we know that new information being added to the genome through mutation is impossible**, this must be wrong.

Explaining speciation is easy for the biblical creationist! When God created each original kind, He gave them a tremendous amount of genetic variability. As an example, a husband and wife have enough genetic variability that they could have 10^{2017} children before they would produce an identical one. That is a 1 with 2017 zeros after it! To even try to get an idea of how big that number is, we can consider our vast universe that contains at least billions of trillions of stars. It is estimated that our entire universe only contains 10^{80} particles. Yet, the chances of a married couple having 2 identical kids (not identical twins which is a result of a single fertilized egg splitting in half) is many times greater than the number of particles in the universe!

Through concepts such as natural selection, genetic drift, and others, the environment causes some genes in the vast genome of the organism to be expressed to the next generation, and others are lost. In a simple example, imagine a pair of dogs that each has a gene for both long fur (L) and short fur (S). Thus, since each of those dogs has an LS combination, they would both have medium length fur. Using basic genetics, their pups could have one of four possible combinations of genes for length of fur: LL (long fur), SS (short fur), LS (medium length fur), or SL (medium-length fur). Now, imagine that they have one hundred puppies with all of the combinations above. Half of the puppies migrate to a hot climate, and the other half migrate to a cold climate. In the hot climate, the long fur (LL) and medium fur puppies (SL, LS) die off before they can reproduce, leaving only the short fur (SS) puppies alive. Now, with the long gene (L) lost from the genome, only short fur (SS) dogs remain to reproduce and make more short fur (SS) dogs. Eventually, a new species of dog can develop. Similarly, in the cold climate, the short fur (SS) and medium fur puppies (SL, LS) die off before they can reproduce, leaving only the long fur (LL) puppies alive. Now, with the short gene (S) lost from the genome, only long fur (LL) dogs remain to reproduce and make more long fur (LL) dogs. Eventually, another new species of dog can develop.

In this oversimplified example, we see how natural selection, genes being "selected" by the environment, plays a role in speciation. However, natural selection *only acts on genes that are already present in the genome.* No new genes or traits arose here; we only see a different combination of genes being expressed. With the vast amount of genetic information that every organism carries, as well as all of the genetic variation present, new species can develop over time.

It is important to remember that the new species are always of the same kind as the parent kind. In the example above, dogs will always produce dogs. No matter how much speciation occurs, no matter how much time passes, the progeny will always be of the dog kind.

Death

The theory of evolution is driven by death. For evolution to be true, death must have existed before sin. It must have always existed. Yet, the Bible clearly teaches that death, both **spiritual and physical death**, was introduced into a perfect world as a *result of Original Sin* (as we will discuss later.) This is shown in many verses, including:

> "By the sweat of your face you shall eat bread, till you return to the ground, for out of it you were taken; for you are dust, and to dust you shall return." (Genesis 3:19)

> "Therefore, just as sin came into the world through one man, and death through sin, and so death spread to all men because all sinned... death reigned from Adam to Moses, even over those whose sinning was not like the transgression of Adam..." (Romans 5:12,14)
> "For the wages of sin is death..." (Romans 6:23a)

In Genesis 3:19 above, God tells Adam that his body came from the ground and that his body would eventually return to the ground *in death* as punishment for sin, just like He promised in Genesis 2. The fact that Adam didn't immediately die was the first example of God's mercy granted to humankind -this is the same mercy that is given to every one of us each time that we sin! We see this exemplified in the following passage:

> "The Lord is not slow to fulfill His promise as some count slowness, but is patient toward you, not wishing that any should perish, but that all should reach repentance."
> (2 peter 3:9)

In the Romans 5 passage above, Paul explains that every person born of man (thus, every person except Jesus Christ) is born with a sin nature passed on from Adam. Thus every person will die because of that sin nature inherited from Adam and the subsequent sins committed throughout life. It is clear by these passages, and many others, that *death is a result of sin.*

> *Thus, we can plainly see that evolution could not have occurred before Adam's sin! Death was only a result of The Fall. According to the Bible, death could not have happened before sin!* **This may be the most important thing to remember in refuting evolutionism!**

It is very clear that evolution is not compatible with the Bible. We will read in chapter 8 why general evolution (macroevolution) is not only a fairy tale, but also is not scientific as it fails getting past the hypothesis stage using the scientific method.

It is vitally important to understand this concept as *the Gospel rests on this fact*. As we will see in chapter 10, Jesus's death on the cross is absolutely necessary to atone for our sins. This is shown in the following verses:

> "Indeed, under the law almost everything is purified with blood, and without the shedding of blood there is no forgiveness of sins." (Hebrews 9:22)

In this verse, we can see that there is no payment for sins without the shedding of human blood. It is important to understand that the shedding of blood here refers to death. The shedding of blood is not sufficient to atone for sin without death accompanying it.

> "Since therefore the children share in flesh and blood, he (Jesus) himself likewise partook of the same things, that through death he might destroy the one who has the power of death, that is, the devil." (Hebrews 2:14)

This verse not only shows that Jesus shed His precious blood and *died* to destroy the one who held the power of death, but also to fulfill the prophecy that was foretold in Genesis 3:15. **God always keeps his promises, and He never changes His standards.**

Chapter Four

The Relevance of Genesis

Does it really matter what a Christian thinks about Genesis? Is it really that important and/or relevant to us today?

Genesis is Real History

We have already established in chapter 1 that the entire Bible is the word of God and is true. Because of this, the first book of the Bible must also be true! However, it is also helpful to look at how some of the New Testament (NT) writers viewed Genesis as literal history.

1. Every author of the NT, and almost every book in the NT, has a direct reference to Genesis 1-11 (creation account, original sin, Adam and Eve, the Flood, Noah and the Ark, etc.)[1] It is understood that the NT writers looked at Genesis as literal history as they referred back to it often. This includes:

>Matthew 19:4
>Mark 10:6
>Luke 3:37-38
>John 1:1-3
>Acts 4:17-26, especially verse 24
>Romans 1:29, 5:12
>Hebrews 1:10, 11:3
>James 3:9 (referring to Genesis 1- humans are made in the image of God)
>1 Peter 4:19
>Jude 1:11,14
>Revelation 10:6

2. Many NT verses rely on an historical account of Genesis. Consider the following verses:

> "Thus it is written, 'The first man Adam became a living being'; the last Adam (Jesus) became a life-giving spirit... The first man was from the earth, a man of dust; the second man is from heaven."
> (1 Corinthians 15:45, 47)

We know from the Bible that Jesus is real. These verses connect the "literal Jesus" with the "literal Adam" - they would make no sense if Adam was not a real individual from history. We are drawn back to the beginning of Genesis where Adam was made into a literal, living being from the dust of the ground.

Remember the genealogies that were mentioned in chapter 1? Several are of particular interest for this discussion. Jude 1:14, speaking of Enoch being the 7th from Adam, is a historical fact from Genesis 5. Luke 3 traces the genealogy of Jesus through His legal father Joseph, all the way back to Adam. This is yet another verse tying a "literal Adam" to a "literal Jesus."

In the Gospels, Jesus's words on earth verified Genesis. He referenced Genesis 8 times, including:

> "He answered, 'Have you not read that he who created them at the beginning made them male and female,' and said, 'Therefore a man shall leave his father and his mother and hold fast to his wife, and the two shall become one flesh'? So they are no longer two but one flesh. What therefore God has joined together, let not man separate." (Matthew 19:4-6)

Jesus is directly quoting from Genesis 1&2, in the very same passage, regarding marriage being between only one man and one woman. In this case, He is speaking against divorce. Jesus also states that Adam and Eve were made at the beginning - which makes sense when you consider that the 6th day of 6000 years is the beginning! If there were millions/billions of years before Adam and Eve, then their creation 6000 years ago would be at the end of the earth's timeline, not the beginning. Consider the following diagram:

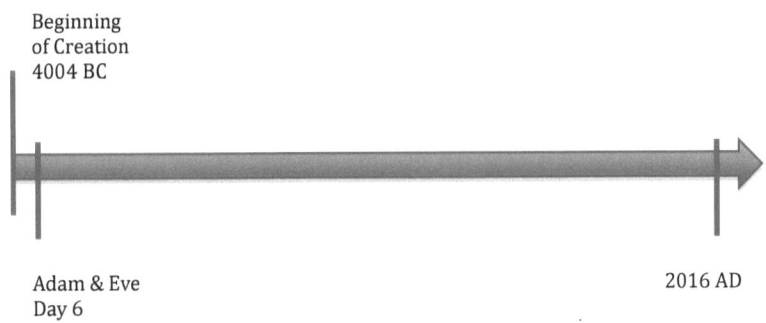

Literal Interpretation: 6,000 Year-old Earth

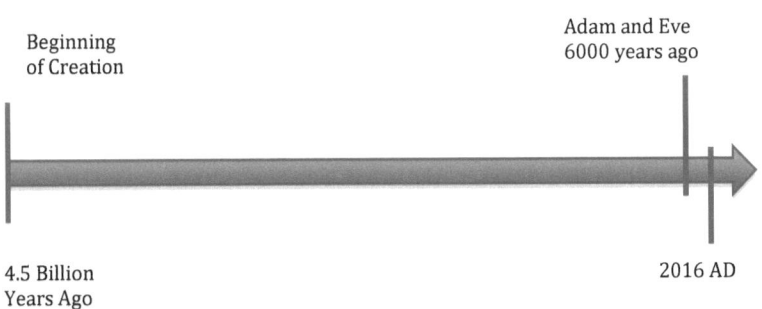

Non-Literal Interpretation: 4.5 Billion Year-old Earth

"Do not think that I have come to abolish the Law or the Prophets; I have not come to abolish but to fulfill them. For truly, I say to you, until heaven and earth pass away, not an iota, not a dot, will pass from the law until all is accomplished.'" (Matthew 5:17-18)

By Jesus saying "the Law or the Prophets," He was referring to the *entire* Old Testament. This is a Jewish way of referring to what Christians call the Old Testament, and it serves as another testimony to a literal Genesis.

> "Do not think that I will accuse you to the Father. There is one who accuses you: Moses, on whom you have set your hope. For if you believed Moses, you would believe me; for he wrote of me. But if you do not believe his writings, how will you believe my words?" (John 5:45-47)

Jesus is verifying Moses in this passage – the author of Genesis through Deuteronomy. Not only does He verify Moses, Jesus says that Moses constantly verifies Him! There are a number of prophecies and "fuzzy pictures" of Jesus contained throughout the OT, starting in Genesis 3:15. We cover this in the next chapter.

Why is the Book of Genesis Important?

When encountering people on the street, the majority of questions that are asked fall into discussions of Genesis 1-11. We cover the most common challenges and how to answer them in chapters 13, 14, and 15. Some of the questions include:

1. The creation account in Genesis 1 and Genesis 2
2. Relationships of incest, starting with whom did Cain marry to be his wife?
3. How did Noah fit all of the animals on the Ark?
4. Where do different languages/races come from?
5. What about the fossil record?
6. If God is so good, then why do bad things happen to good people?
7. Why do kids get cancer? (This is one of many variations of #6 above)
8. Who made God?

All of these challenges require a correct understanding (literal interpretation) of Genesis. We will see throughout this book that if you believe that the Bible is 100% true, a young earth interpretation of Genesis is the **only** correct one. Every other idea for origins (evolution, theistic evolution, and other old earth ideas) will conflict with many other parts of scripture.

The most common question asked of Christians by unbelievers today is some variation of #6 above. To be able to answer these questions Biblically, we need to be able to properly explain why tragedy and suffering exist in this world. When Genesis 2 and Genesis 3 are taken literally, we can explain how original sin corrupted this world that God created perfectly in the beginning. God isn't the author of death, disease, suffering, etc. – through Adam and Eve, **we are the authors of everything bad**.

The Exodus Because of Genesis

Research conducted by respected pollster George Barna[2] shows that nearly 2/3 of all young adults who grow up in Christian homes, and attend church regularly, become "spiritually disengaged" from the church during their college-age years. No more attending church, no more praying, and no more reading their Bibles.

Britt Beemer, of America's Research Group, studied this issue further at the urging of Ken Ham from Answers in Genesis. The book, *Already Gone*, by Brit Beemer and Ken Ham[3], displays this startling research reported by Britt and his America's Research Group. The answers from these 20-somethings are shocking, as shown in the following two examples:

Of these 20 to 29-year-old evangelicals who attended church regularly but no longer do so:

95% of them attended church regularly during their elementary and middle school years 55% attended church regularly during high school

Of the thousand, only 11% were still going to church during their early college years

Young Adults who no longer believe that all of the accounts and stories in the Bible are true:

39.8% first had doubts during middle school
43.7% first had doubts during high school
10.6% first had doubts during college

The church has always assumed that these young adults leave college during the college years where newfound freedoms are to be had. Yet, the results of this study show something completely different: they were already lost long before they went to college! **A large percentage of the doubts shown above are regarding- you guessed it- the book of Genesis.**

We must, however, be cognizant of something. The kids that "left" the church were not saved to begin with. In a quote from Grace to You Ministry, "The issue of determining who is a Christian becomes difficult when considering people who claim to believe. Many who profess to believe in Christ may act like Christians to a certain degree, but turn out to be impostors or are simply misinformed about the nature of salvation. Either way, it becomes obvious that they did not really know the truth. For example, (the following verse) identifies some people who claimed to be Christians but eventually left the fellowship:"[4]

> "They went out from us, but they were not of us; for if they had been of us, they would have continued with us. But they went out, that it might become plain that they all are not of us." (1 John 2:19)

As we see in the Grace to You (GTY) commentary, parents, pastors, and Sunday school teachers did not provide the answers for the children, starting with the nature of salvation.

The Education of our Kids

Hitler once quoted, "He alone, who owns the youth, gains the future."[5] He said this in regards to preparing children to join his Nazi regime.

Karl Marx, the "Father of Communism" and the author of *The Communist Manifesto,* stated, "The education of all children, from the moment that they can get along without a mother's care, shall be in state institutions at state expense."[6]

In light of these quotes, think about these stats:

1. There are 8760 hours in a year.
2. Kids sleep for 2920 hours per year, 1/3 of our time, assuming 8 hours of sleep per night.
3. School takes up over 1100 hours per year[7]
4. Time spending watching TV equals over 1000 hours per year[8]
5. Playing computer games and socializing over 400 hours per year[8]

These activities alone take up over 5400 hours in a year, or 62% of a child's total time. This equates to nearly 15 out of every 24 hours! When one adds in video games, time with friends on the phone, and social media use, how much time is left for even the most well intentioned parents to train up his/her child?

The reality is that the decaying of society that we see today can largely be traced to the indoctrination of our children by these unbiblical sources. John Dewey and his influence on the education system at the beginning of the 20th century has been the major influence in the decay of morality in society over the last fifty years. The progressive movement, that he helped spurn, has been indoctrinating our children in the schools, on television, and in society in general. In our public schools over the years, we have seen a major increase in rates of premarital sex, pregnancy, abortion, drug abuse, alcohol abuse, and many other negative indicators. The progressive movement has been shaping the worldview of our children in a major way. *They are being brainwashed, enslaved, and indoctrinated in anything and everything but the Christian worldview. In fact, it is an anti-Christian worldview.*

An Authority Issue

As stated earlier, the issue of whether Genesis is to be taken literally or not is one of *authority*. Genesis, as well as the rest of the Bible, clearly infers that the earth is only about 6000 years old and the modern teaching of evolution is not in the Bible. The *Law of Christ*, what God tells us in how He created, and the *law of evolution*, what some scientists tells us about how we were created, are in direct opposition. We cannot amalgamate the two, and they cannot both be right. However, this issue of authority (God's Word vs. man's opinion) runs even deeper than most people realize.

Think about some of the *crazy* things that we believe as Christians, many of which conflict with "science." On the surface, the following examples are no different than what some scientists believe, such as nothing turning into everything and frogs turning into princes! Yet, we know these to be true because God tells us so:

> A talking donkey (Numbers 22:28)
> An axe head that floats (2 Kings 6)
> A virgin birth (Matthew 1:1-18, many others)
> Walking on water (Matthew 14).
> Turning water into wine (John 2: 1-11)
> Bringing dead people back to life (John 11:1-44)
> Jesus raising Himself from the dead, as will be
>   discussed in chapter 6 (John 2:19-22).

When our starting point is God's Word, our authority is the Bible *alone*. Consider the following passages:

> "For the foolishness of God is wiser than men, and the weakness of God is stronger than men." (1 Corinthians 1:25)

> "For although they knew God, they did not honor him as God or give thanks to him, but they became futile in their thinking, and their foolish hearts were darkened. Claiming to be wise, they became fools." (Romans 1:21-22)

We are created in *Imago Dei*, which means in the image of God (not from a common ancestor shared with monkeys) and He is the *absolute authority on everything,* including morality. When our starting point is man's word (opinion), authority becomes *relative*. If evolution were true, and we are just the result of rearranged pond scum from billions of years ago, morality is *relative*. There is no absolute standard because we are just "molecules in motion." Besides, the "morality" that is necessary for "natural selection/ survival of the fittest" does not make any sense- killing weaker people should be a good thing in an evolutionary worldview! Morals do not arise from random chemical reactions and the killing of the weak as time passes. **The reality is that man lives by morals because He is made in the image of God.** Random chemical reactions cannot produce absolute, immoral, universal entities like morality. In addition, the Bible also shows that God wrote His moral law on everyone's heart.

Quoted from Todd Friel's book, *Jesus Unmasked*, "Every person on the planet has a little courtroom in their brain called 'the conscience.' Your conscience is that voice in your noggin that nags you when you have done something wrong, and everyone has one. Evolution simply can't explain why every culture shares the same basic set of values... Where did the conscience come from? Certainly not from Charles Darwin."[9]

The conscience that is often portrayed in cartoons -an angel on one shoulder and Satan on the other- was implanted into us by God!

> "For when Gentiles, who do not have the law, by nature do what the law requires, they are a law to themselves, even though they do not have the law. They show that the

work of the law is written on their hearts, while their conscience also bears witness, and their conflicting thoughts accuse or even excuse them." (Romans 2:14-15)

We look at moral issues today, such as murder, abortion, homosexuality, school killings, etc., and we think that they are the problems. In reality, they are just the *symptoms* of the problem. The problem is clearly written at the end of the book of Judges. After reading Judges 18-21, one can read what the real problem is in the very last verse:

"In those days there was no king in Israel: every man did that which was right in his own eyes." (Judges 21:25)

The problem today is that the King of Kings, the Lord of Lords, is being removed from everything in this country. When the truth about Him is suppressed (Romans 1:18-20), absolute morality also gets suppressed. (This is explained further in chapter 7) Jesus Christ is the only solution! When the problem gets fixed, the symptoms start to disappear! Why? *Because He is our absolute authority!*

A true understanding of God is dependent upon having a correct understanding of His Word. *And the foundation of His Word is in the beginning of the book of Genesis.*

Genesis: The Foundation

Genesis, which is the first of the sixty-six books in the Bible, is comprised of fifty chapters. What is not often realized is that Genesis 1-11 covers over 2000 years of time, which is 1/3 of all of earth's 6000-year history! Through the first 11 chapters, we get the Creation account, the Fall, the Flood (and Ark), and the Tower of Babel, which explains such things as the formation of different people groups, languages, and false religions throughout the world.

Many Biblical doctrines also have their foundations in Genesis:

1. Marriage (including the fact it is only between one man and one woman)
2. Laws
3. Standards
4. The meaning of life
5. Animal sacrifice (Leading up to the ultimate sacrifice, Jesus Christ)
6. Concept of a week (Days, Months and Years come from astronomy. A week comes only from the creation account and is repeated by God in Exodus 20)
7. The necessity of clothing

The Bible is meant to be read and understood in a plain or straightforward manner. "Reading the Bible 'plainly' means understanding that literal history is literal history, metaphors are metaphors, poetry is poetry, etc. The Bible is written in many different literary styles and should be read accordingly."[10] The first eleven chapters of Genesis are clearly written as a historical narrative, as we have addressed in chapter 14 of this book. It records actual historical events with essential doctrines that result from those events. **The book of Genesis must be taken literally!**

Chapter Five

Creation, The Fall, and The Promise

Genesis: The First Three Chapters

The first three chapters of Genesis cover: God's perfect creation (Genesis 1 and Genesis 2), The Fall (Genesis 3), The Curse over all creation (Genesis 3), and God's plan of redemption, the promise of the Savior (Genesis 3:15)

In Genesis 1, God created everything over six 24-hour literal days. (This will be discussed in chapter 12) Notice that I did not say "solar days." The sun was not created until day 4. Many people wrongly object to God creating in six 24-hour literal days because there was no sun present in days 1-3. Therefore, they believe, that the creation account cannot be literal. However, this is not actually a problem because a day is defined as one rotation of the earth on its axis. It has nothing to do with the sun- it just so happens that the sun *appears* to rise and set each day. (This will be discussed in chapter 15). At the end of day 6, His Creation was deemed perfect:

> "And God saw everything that he had made, and, behold, it was very good. And there was evening and there was morning, the sixth day." (Genesis 1:31)

The word "good" is used six times in Genesis 1. At the end of His creating on Day 6, and looking over His entire creation, God said "good" for the 7th time: "... It was very good." This *perfect* creation is also described in many verses of the Bible- in reference to God creating a new Garden of Eden at the consummation- including the following verses from Revelation 21:

"Then I saw a new heaven and a new earth, for the first heaven and the first earth had passed away, and the sea was no more. And I saw the holy city, new Jerusalem, coming down out of heaven from God, prepared as a bride adorned for her husband. And I heard a loud voice from the throne saying, "Behold, the dwelling place of God is with man. He will dwell with them, and they will be his people, and God himself will be with them as their God. He will wipe away every tear from their eyes, and death shall be no more, neither shall there be mourning, nor crying, nor pain anymore, for the former things have passed away." And he who was seated on the throne said, "Behold, I am making all things new." Also he said, "Write this down, for these words are trustworthy and true." (Revelation 21:1-5)

Genesis 2, a detailed account of Day 6, tells us of the one main command given to Adam. (It needs to be noted here that the command was not given to Eve, as she was not yet created.)

"And the Lord God commanded the man, saying, 'You may surely eat of every tree of the garden, but of the tree of the knowledge of good and evil you shall not eat, for in the day that you eat of it you shall surely die.'" (Genesis 2:16-17)

It is important to understand that this command did not mean that Adam would *immediately* drop dead physically. What it did mean is that Adam would die both spiritually (immediate) and physically (later). We will see Biblical proof of these statements later in the chapter.

In Genesis 3, we get a correct understanding of why bad things are in the world - it comes as a result of "The Fall." We see this play out in the first several verses.

"Now the serpent was more crafty than any other beast of the field that the Lord God had made. He said to the woman, 'Did

God actually say, 'You shall not eat of any tree in the garden'?" (Genesis 3:1)

This verse gives us the "first sighting" of Satan in the Bible. He came in the form of a serpent and challenged God by asking Eve a question. Notice how Satan used God's Word (the command given to Adam in Genesis 2) but changed it slightly. This is reminiscent of the false teachers of today, like the prosperity gospel teachers, Word of Faith (WOF) preachers, and those of the New Apostolic Reformation (NAR).

"And the woman said unto the serpent, 'We may eat of the fruit of the trees in the garden, but God said, 'You shall not eat of the fruit of the tree that is in the midst of the garden, neither shall you touch it, lest you die'.'" (Genesis 3:2-3)

In these two verses, we see Eve's response to the serpent. She also used God's Word to respond - and also got it wrong. As stated earlier, *the original command was given to Adam only.* It is interesting to note that something went awry in the communication of God's Word from Adam to Eve. One of the following things must have happened, though we cannot know which based on Scripture: Adam did not teach his wife correctly regarding God's Word, Eve just didn't listen well to correct teaching from Adam and didn't learn it, or Eve knew the correct teaching and didn't articulate it correctly to the Serpent.

Next, we see the answer given by the Serpent:

"But the serpent said to the woman, 'You will not surely die. For God knows that when you eat of it your eyes will be opened, and you will be like God, knowing good and evil'." (Genesis 3:4-5)

The serpent challenged God and His Word in his response to Eve. He also told Eve that she and Adam could be like God Himself, knowing good and evil. What shall Eve do when given the ability to be like God? How did she handle the temptation? It didn't take long to find out....

> "So when the woman saw that the tree was good for food, and that it was a delight to the eyes, and that the tree was to be desired to make one wise, she took of its fruit and ate, and she also gave some to her husband who was with her, and he ate. Then the eyes of both were opened, and they knew that they were naked. And they sewed fig leaves together and made themselves loincloths." (Genesis 3:6-7)

As their eyes were suddenly opened, the world, starting with them, was now exposed to evil. They immediately died *spiritually*. The first thing that they noticed was that they were naked! They were naked all along, but now they knew the evil part of their nakedness. This is exactly why Genesis 2 ends with the following verse that seems to not really flow with the rest of the chapter:

> "And the man and his wife were both naked and were not ashamed." (Genesis 2:25)

While this verse seems to just be thrown at the end of Genesis 2 and not belong there, it makes sense in the context of Genesis 3!

Their reaction was to immediately sew fig leaves to cover their nakedness and shame, and they attempted to hide from God. This probably sounds familiar, as we humans tend to attempt to hide when we sin. What happens next is God confronting Adam with his sin.

> "And they heard the sound of the Lord God walking in the garden in the cool of the day, and the man and his wife hid themselves from the presence of the Lord God among the trees

of the garden. But the Lord God called to the man and said to him, 'Where are you?' And he said, 'I heard the sound of you in the garden, and I was afraid, because I was naked, and I hid myself.' He said, 'Who told you that you were naked? Have you eaten of the tree of which I commanded you not to eat'?" (Genesis 3:8-11)

Spiritual death happened immediately to Adam and Eve, and the promise of an eventual physical death would soon follow. The whole world would start to decay as a result of this original sin. We will see the Biblical proof of this next.

Genesis: The Fall and the Promise

Because of this original sin, we now see all the bad entering the world, including death, disease, suffering, extinction, thorns and thistles, and so on. *All* of creation was affected:

> " For the creation was subjected to futility, not willingly, but because of him who subjected it, in hope that the creation itself will be set free from its bondage to corruption and obtain the freedom of the glory of the children of God. For we know that the whole creation has been groaning together in the pains of childbirth until now. And not only the creation, but we ourselves, who have the firstfruits of the Spirit, groan inwardly as we wait eagerly for adoption as sons, the redemption of our bodies." (Romans 8:20-23)

The next thing that we see is the blame game: Adam blamed God for giving him Eve, and Eve pointed her finger at the serpent for his deception.

"The man said, 'The woman whom you gave to be with me, she gave me fruit of the tree, and I ate.' Then the Lord God said to the woman, 'What is this that you have done?' The woman said, 'The serpent deceived me, and I ate'." (Genesis 3:12-13)

God then starts to curse His creation, starting with the serpent:

"The Lord God said to the serpent, 'Because you have done this, cursed are you above all livestock and above all beasts of the field; on your belly you shall go, and dust you shall eat all the days of your life'." (Genesis 3:14)

God continues to curse His creation, now speaking about humanity:

1. He multiplies pain during childbirth. (Genesis 3:16)

2. He shows the struggle that will take place between husband and wife in marriage, ruining the original perfect harmony. The woman will try to gain power and usurp authority over her husband, but the husband will attempt to dominate her. (Genesis 3:16)

3. The ground was cursed: thorns and thistles would now appear in the garden between the plants and trees that they would eat from. Work, which God did intend for Adam in Genesis 2, would now become hard and laborious for them. (Genesis 3:17-19)

The last part of the curse is shown in the following verses. It is clear that physical death would be the *end result* of original sin.

"By the sweat of your face you shall eat bread, till you return to the ground, for out of it you were taken; for you are dust, and to dust you shall return… Then the Lord God said, 'Behold, the man has become like one of us in knowing good and evil. Now, lest he reach out his hand and take also of the tree of life and eat, and live forever—' therefore the Lord God sent him out from the garden of Eden to work the ground from which he was taken. He drove out the man, and at the east of the Garden of Eden he placed the cherubim and a flaming sword that turned every way to guard the way to the tree of life." (Genesis 3:19,22-24)

We have seen how God cursed His creation as a result of Original Sin, and the bookend verses of this are Genesis 3:7 (immediate spiritual death) and Genesis 3:19 (an impending, certain physical death). However, in the middle of these verses is something so glorious! God gives us the promise of the coming Savior - the first prophecy of the Lord Jesus Christ:

"And I will put enmity between thee and the woman, and between thy seed and her seed; it shall bruise thy head, and thou shalt bruise his heel." (Genesis 3:15 KJV)

We see here that the Savior would come from the seed of the woman, which is what makes the genealogies so important throughout the Bible!

Now, with the shame of nakedness and the promise of the Savior, yet another important concept is born here in Genesis 3, which lasts the entirety of the Old Testament - animal sacrifice (bloodshed) to temporarily cover sin.

"And the Lord God made for Adam and for his wife garments of skins and clothed them." (Genesis 3:21)

While Adam and Eve covered their shame with fig leaves (Genesis 3:7), God demanded blood. The first animal sacrifice occurred in the Garden of Eden, with God Himself killing the first animal and using the skins of the bloodshed animal to cover their sin.

This pattern is repeated throughout the Old Testament, as the Israelites were instructed to slay a first-born, unblemished animal to be a temporary cover for their sin. These sacrifices foreshadowed the perfect and complete sacrifice of Jesus Christ – the unblemished, first-born, Lamb of God- who took away the sins of the world. We know from Hebrews 9:22 that, "…and without shedding of blood is no remission (of sins)."

It is easy to see why a literal interpretation of the beginning of Genesis is vital to a correct understanding of the rest of the Bible. God's plan of redemption is based on what happened in Genesis. Considering all of this, something else should be very apparent. The Bible teaches that death is only a result of sin. Because death is the result of sin, death could not have been present before Adam and Eve's original sin. Therefore, evolution never happened, as it requires death to always have been present! This will be covered in a few of the chapters of this book.

One more thing: I have a question. Based on your knowledge from this book so far,

"Why did Jesus have to die?"
(Hint: Review Genesis 2 and 3.)

A pastor answered my question, in this way, a few months ago: "I know this is a trick question and all…"

My retort: "I can assure you, Pastor, that this is not a trick question." Why did Jesus have to die?

After many seconds of silence, I offered him some help: I walked him through Genesis 2 and Genesis 3 for the answer. It appears later in the book!

Chapter Six

The Gospel

The power of God unto salvation

Is the "funny Pastor" the power of God unto salvation?
-No!

Is the "well-conceived argument" the power of God unto salvation?
-No!

Is the "Jesus loves you" declaration the power of God unto salvation?
-No!

Is the "God has a wonderful plan for your life" the power of God unto salvation?
-No!

"...*The Gospel*...is the power of God unto salvation to every one that believeth." (Romans 1:16 KJV)

So, what is the Gospel?

The Gospel

Imputation. That's the gospel in its shortest possible form. What does that word mean? Simply defined by Webster's dictionary, it means, "To credit to a person." The KJV Dictionary states it as follows: The act of imputing or charging; attribution; generally in an ill sense; as the imputation of crimes of faults to the true authors of them."[2]

Why would imputation explain the Gospel? When Jesus died on that cross, His righteousness was imputed to us, and our sin was credited to Him. He paid the fine that we owe to God for our sins- we would not have to pay it on our own. The fine is eternal because our sin is an offense to an eternal God. To pay our fine and give us everlasting life, Jesus had to be fully God (eternal) and fully man. He had to be fully man because human bloodshed is required to pay the penalty that each of us owes to God. He had to be fully God in order to give us everlasting life in Him. Our salvation came at the highest cost. One that even the most lustrous of writers could never adequately describe.

In 1 Corinthians 15, Paul is reviewing the Gospel, re-affirming his apostleship and talking about the eternal life that we as Christians can look forward to. He is reviewing the Gospel to remind the church at Corinth how they are saved. This is what we will be covering now.

The passage I will be starting with, and continually referring to, is 1 Corinthians 15:3-6,

> Verse 3. "For I delivered to you as of first importance what I also received: that Christ died for our sins in accordance with the Scriptures,"
>
> Verse 4. "That he was buried, that he was raised on the third day in accordance with the Scriptures,"
>
> Verse 5. "And that he appeared to Cephas, then to the twelve."
>
> Verse 6. "Then he appeared to more than five hundred brothers at one time, most of whom are still alive, though some have fallen asleep."

Paul, whom God inspired to write this book and a large portion of the New Testament, has an amazing story. He first appears in the Bible in Acts 7, at the murder of Stephen. He was a part of the gathered crowd, and he watched everyone's garments, as well as giving the religious support to stone him because he was a Pharisee. This also gave them the ability to stone Stephen to death. Before his conversion, Saul (Paul) was a Pharisee- a religious leader of the day. As he stood there watching their garments, he was giving his full support as a leader in this murder. This stoning came up as a flash of rage. If you remember from when Jesus was on trial, the Jews said to Pilate in John:

"It is not lawful for us to put anyone to death." (John 18:31)

The governor at that time would let certain things go under the radar to keep from causing any riots. This is similar to what was on the verge of happening at the trial and sentencing of Jesus in John 19, Luke 23, Mark 15 and Matthew 27.

After the public slaughter of Stephen, there was a mass outcry against Christians, as they were known as followers of the way at that time. Saul went into people's houses and dragged them off to prison. During this time, Saul went to the head of the Pharisees to get authorization to pursue other Christians to try to destroy the church of Christ. This guy was serious about his hatred for Christians because he thought it was not only idolatry, but blasphemy as well. He was willing to travel to cleanse the earth of Christians thinking he was doing something good. This is exactly what Jesus prophesied about in John 16:

"They will put you out of the synagogues. Indeed, the hour is coming when whoever kills you will think he is offering service to God." (John 16:2)

Soon after receiving this letter of authorization from the religious leaders, he headed out of town to start trying to persecute the church. While he traveled along the road to Damascus, Saul was surrounded by a light and fell on the ground. He heard Jesus speaking to him- confronting him with the charges He had against Saul. The fact is that Saul was persecuting Him (meaning His church, which is earlier described by Paul in 1 Corinthians 12). Saul had no idea who was speaking to him, but, when he asked, Jesus told him who He was. Jesus then instructed him to go into the city and he will be told what to do. Saul, by this point, was actually physically blinded from the experience.

Afterward, the Lord came to Ananias in a vision, telling him to go to where Saul was and lay his hands on him, praying that he might regain his sight. He told Ananias that Saul is a chosen instrument to carry His name before the Gentiles and kings and the children of Israel. He went to Saul and prayed for him. After being healed from his blindness and receiving the Holy Spirit, he was baptized. A few days later, Paul (formerly Saul) started proclaiming the Gospel in the synagogues, and growing daily he continued to confound the Jews and continued to prove that Jesus was the Christ.

All of this background is for a reason. It is to show you the changing power of the Gospel of Jesus Christ. When we first saw Saul, he was supporting the murder of Christians, and he helped drag people to prison for believing that Jesus was the Messiah. Messiah means, "the anointed one of God." After he met Jesus and was saved, he turned into the most vocal of all the Christians by going into the synagogues proving that Jesus is Lord! There wasn't anyone who could argue with him, and no one understood what happened to him.

1 Corinthians 15:3

So how does the knowledge above play in with the 1 Corinthians passage from above? Let's start by looking at verse 3. It is what he *received*. He received the Gospel and was saved. He was radically changed and made into a new creature. A heart of stone was removed, and a heart of flesh was put into its place (Ezekiel 36:26). He was born again, and he was completely different! This also illustrates that you cannot come into the presence of God and leave the same person. ***You will, at your core, be completely different.***

In verse 3b, he then stated, "According to the scriptures." At that time, the only scriptures they had were the books that made up the Old Testament. So, let us start looking at some of those scriptures. There are over 300 prophecies about Jesus's coming in the Old Testament. The very first one is in Genesis 3, spoken by God almost immediately after Adam and Eve first sinned:

> "And I will put enmity between thee and the woman, and between thy seed and her seed; it shall bruise thy head, and thou shalt bruise his heel. (Genesis 3:15 KJV)

This prophecy is a great promise of the Messiah to come. Some of the scriptures that speak of Jesus' death can be found in Isaiah:

> "As many were astonished at you- his appearance was so marred, beyond human semblance, and his form beyond that of the children of mankind." (Isaiah 52:14)

He was beaten so badly that you couldn't easily recognize Him:

> "Surely he has bore our griefs and carried our sorrows; yet we esteemed him stricken, smitten by God, and afflicted. But he was pierced for our transgressions; he was crushed for our iniquities; upon him was the chastisement that brought us peace, and with his wounds we are healed. All we like sheep have gone astray; we have turned- every one-to his own way; and the LORD has laid on him the iniquity of us all."
> (Isaiah 53:4-6)

This is the absolute clearest example you can read. I have read this verse to strangers who visited the church off and on while growing up or even those who didn't go to church at all, and they all knew and recognized whom this was describing. It is unmistakable whom this passage is talking about. Yet there are even more than that:

> "Out of the anguish of his soul he shall see and be satisfied; by his knowledge shall the righteous one, my servant, make many to be accounted righteous, and he shall bear their iniquities." (Isaiah 53:11)

He makes us to be counted righteous, and He bears all of our iniquities. We are made right in God's eyes. Jesus paid the debt we owe for breaking God's law.

1 Corinthians 15:4

In 1 Corinthians 15:4, we see the importance of the burial and resurrection as part of the Gospel message.

There are many scriptures, in both the Old Testament and New Testament, which speak of Jesus being dead and raising three days later, conquering death.

> "And they made his grave with the wicked and with a rich man in his death, although he had done no violence, and there was no deceit in his mouth." (Isaiah 53:9)

> "It will be counted to us who believe in him who raised from the dead Jesus our Lord, who was delivered up for our trespasses and raised for our justification."
> (Romans 4:24-25)

> "But God shows his love for us in that while we were still sinners, Christ died for us." (Romans 5:8)

> "For you will not abandon my soul to Sheol, or let your holy one see corruption." (Psalm 16:10)

A reference in the New Testament of Jesus's death and resurrection also ties into the Old Testament. In Matthew, Jesus refers to the sign of Jonah - this is a prophecy that Jesus spoke about Himself only being dead for 3 days and then rising.

> "So will the Son of Man be three days and three nights in the heart of the earth." (Matthew 12:40b)

It is important to note that Jesus's resurrection is vital to the Gospel message:

> "Now if Christ is proclaimed as raised from the dead, how can some of you say that there is no resurrection of the dead? But if there is no resurrection of the dead, then not even Christ has been raised. And if Christ has not been raised, then our preaching is in vain and your faith is in vain. We are even found to be misrepresenting God, because we testified about God that he raised Christ, whom he did not raise if it is true that the dead are not raised. For if the dead are not raised, not even Christ has been raised. And if Christ has not been raised, your faith is futile and you are still in your sins. Then those also who have fallen asleep in Christ have perished. If in Christ we have hope in this life only, we are of all people most to be pitied." (1 Corinthians 15:12-19)

1 Corinthians 15:5

In verse 5, Paul is speaking about when Jesus visited the disciples and others after the resurrection, which can be found in all of the Gospel accounts -Matthew 28:16; Mark 16:14; Luke 24:36; John 20:26- as well as Acts 1. The Bible tells us one of the proofs of the Resurrection here- that there were over 500 eyewitness accounts of Jesus walking the earth after He was raised. We will speak more about this in chapter 9.

I pray that something is evident while reading the prophecies and the fulfillment of them. God keeps His word and He knows what will happen in the future! It's one of His incommunicable attributes- characteristics that we don't share in.

All Have Sinned

As stated earlier, imputation sums up the Gospel. Recognize it or not, we have all sinned and our sin, before salvation is imputed to us which leaves us with a fine we cannot pay. Again, it is only a cost Jesus could pay on our behalf – it makes His imputing His righteousness onto us so important. What is sin? It's a transgression or a breaking of God's law:

> "Everyone who makes a practice of sinning also practices lawlessness; sin is lawlessness." (1 John 3:4)

This law, in its simplest form, can be found in Exodus 20:1-17. *It is also known as the 10 commandments.* As it says in Romans 3:

> "For all have sinned and fall short of the Glory of God." (Romans 3:23)

Not a single person has kept the law perfectly, **except for Jesus**. Think about that for a second- not even a single sinful thought crossed his mind. How many sinful thoughts dance between our ears every hour that we live? More than I would like to admit!

If someone were to say, "Well, I have kept a few of the Commandments," then we should see what the Lord says through James:

> "For whoever keeps the whole law but fails in one point has become accountable for all of it." (James 2:10)

So, if we break even one commandment at any point in our life, then we have broken them all.

The Gospel Call

What importance does this have for us? An *eternal one*. If we are not found in Christ before the very moment we die or when He returns (whichever comes first), the wrath of God abides on us. As it says in John 3:

> "Whoever believes in the Son has eternal life; whoever does not obey the Son shall not see life, but the wrath of God remains on him." (John 3:36)

This is an amazing offer He gives to us. Jesus descended and became like us in flesh *but did not sin*. Yet, in His love for us, He paid the price that we owe to God on the Cross. He took the wrath of God on the Cross, bearing the weight of our sin:

> "That is, in Christ God was reconciling the world to himself, not counting their trespasses against them, and entrusting to us the message of reconciliation… For our sake he made him to be sin who knew no sin, so that in him we might become the righteousness of God." (2 Corinthians 5: 19,21)

Martin Luther considered this the "Great Exchange." Christ, the Spotless Lamb, bore *our* sins on the Cross, and we receive His righteousness in exchange! The grace and mercy we do not deserve! Yet, we are offered this olive branch of peace. As a result of this, we are called to do as Jesus said in Mark 1:

> "The time is fulfilled, and the kingdom of God is at hand; repent and believe in the Gospel." (Mark 1:15)

Repentance is a change of mind and heart regarding sin. God grants you the ability to do so. It is a complete 180-degree turn. Believing the Gospel is trusting that Jesus died for you and rose again according to scriptures, proving His sacrifice was sufficient for the payment for our sins. It is putting all of your faith in Jesus- in Him alone- for your salvation. *There is no other way, contrary to popular belief.* Jesus said it Himself in John when He said,

> "I am the way, and the truth, and the life. No one comes to the Father except through me." (John 14:6)

That is a final and exclusive statement only He could make. This is the Gospel. You have sinned continuously, but God already paid the price and offers you salvation - **for free.** You can't do anything to earn it, no matter what anyone says:

> "For by grace you have been saved through faith. And this is not your own doing; it is the gift of God, not a result of works, so that no one may boast." (Ephesians 2:8-9)

A lawbreaker, whether he be a murderer, thief, rapist, speed limit breaker, or anything else, cannot stand in front of a fair and uncorrupt judge and appeal to good works to get out of the deserved punishment. How much more must our Creator, Maker of heaven and earth, the Holy, Righteous, and Just judge, punish us for breaking his Law? There is no amount of good works that we could do to erase the debt we owe. Thank you, Jesus, for paying the fine on our behalf by being the propitiation of our sins!

Chapter Seven

An Introduction to Presuppositional Apologetics

Worldview and Presuppositions- A Review

As we learned in chapter 1, a person's worldview is made up of his set of presuppositions. These presuppositions are certain beliefs that are assumed from the beginning (things that we *pre-suppose* to be true). Every person has a worldview, and every part of a conversation will be interpreted through his own worldview.

The Christian Worldview in its basic form (as we learned in chapter 1) is that: God exists, the Bible is His Word, and, therefore, the Bible is completely true. **Basically, read your Bible and believe what it says!** The rest of this chapter will be devoted to learning why *one cannot account for anything (knowledge, truth, etc.) without the one, true, creator God (of the Bible) as our starting point.* Without God, one cannot make sense of anything.

A Valid Worldview

For a worldview to be valid, it must pass a 3-item test, called the *AIP test* (as coined by Dr. Jason Lisle in his book, *The Ultimate Proof of Creation):*

1. It cannot be arbitrary (without justification).
2. It cannot be inconsistent.
3. It must satisfy the preconditions of intelligibility (such as having a basis for absolute knowledge, and absolute morality)

It is also important to bring up something that we will be establishing later in this chapter, the 2nd Law of Logic, which is the Law of Non-Contradiction.[3] This law states, "... that A cannot be both A and not A at the same time and in the same sense. In other words, something (a statement) cannot be both true and false at the same time and in the same way." This is important because, "Truth is not self-contradictory."[3] This concept is important to understand, as we learn next why a valid worldview cannot be arbitrary or inconsistent.

Is it Arbitrary?

For a worldview to be valid, it first must not be *arbitrary*. Merriam-Webster defines it this way: "Depending on individual *discretion* and not fixed by law... based on or determined by individual preference or convenience rather than by necessity or the *intrinsic* nature of something."[2]

Thus, a valid worldview must be based on correct reason and/or evidence. It cannot be discretionary, or left to one's judgment, choice, or preference. Examples of arbitrary beliefs would be anything that someone says without any support or evidence, such as "I will reincarnate into a flower when I die." Why is this important? Because, if a worldview is left up to one person's discretion, then it is possible for someone else to have a belief that contradicts it. The first person's worldview could not be valid at that point. When a worldview is not arbitrary, it cannot actually be contradicted. Thus, a worldview can be valid only when it is non-arbitrary.

Is it Inconsistent?

Second, for a worldview to be valid, it must also *be consistent*. It must be able to support itself and not have any things that conflict within it. Why is this important? A worldview cannot be trusted to be valid if it conflicts with itself! An example of this would be any of the false religions such as Islam. Islam teaches that the Bible is the Word of God and that God's Word cannot change (be corrupted) (Qu'ran,

Surah 6). Yet, Islam also teaches that followers corrupted the Scriptures.

In addition, regarding the doctrine of the Trinity, the Qu'ran teaches that the three persons of the Trinity are the Father, the Son, and Mother (Mary), while the Bible clearly teaches that the Trinity consists of the Father, the Son, and the Holy Spirit. In both cases, *Islam teaches two things that cannot both be true; Islam is inconsistent and, thus, cannot be a valid worldview.*

Does it Satisfy the Preconditions of Intelligibility?

Lastly, for a worldview to be valid, it must satisfy the preconditions of intelligibility. These include:

- A basis for knowledge
- Uniformity of nature
- Absolute Morality (Isaiah 45:19)
- Reliability of senses, memory, ... (Proverbs 20:12)
- Laws of Logic

A Basis for Knowledge

It can be easily demonstrated that one cannot know anything *for certain* unless he is omniscient (knows everything). Why would this be the case? If a person, let's call him Bob, is not all-knowing, then it follows that there are things that Bob doesn't know. It would be entirely possible that what Bob doesn't know could render false everything that Bob thinks he knows to be true because of his limited knowledge and understanding. Thus, once Bob admits that he is not omniscient, you can show that he cannot know anything for certain. Anything he thinks he knows could be refuted by what he does not know!

Most likely, Bob will acknowledge this but then try and turn it back on you. He will tell you that you are in the same position as he is. You can kindly point out that you are not in the same position as him because of your worldview. In the Christian worldview, God is omniscient, and He has revealed some truth to us that we may know it for certain! As Christians, we have a basis for knowledge, knowing things for certain, because of the fact that we can know things for certain on account of our omniscient God!

This can be explained in the following dialogue that is taught by Creation Today's Eric Hovind, which is something that is easy to use while witnessing. When a non-Christian makes some sort of knowledge claim...

> Christian: "How much knowledge of this entire universe, made up of at least billions of trillions of stars that we know of, do you have?
>
> Non-Christian: "Not much at all."
>
> Christian: "Well, let's assume that you have 1% of all knowledge."
>
> Non-Christian: "That is very generous of you, but I don't have anywhere near that amount."
>
> Christian: "You are right. Neither of us have that much. But let's just assume that you have 1% of all knowledge of the universe. Is it possible, that of the 99% of all knowledge that you don't know, it could refute the entire 1% you think you know?"
>
> Non-Christian: "Well, yes." (The non-Christian realizes quickly that he must answer in the affirmative if he is to be intellectually honest with you.)
>
> Christian: "So, then you can't possibly know anything *for certain*. One must possess all knowledge, be omniscient, to be able to know anything *for certain*.
>
> Non-Christian: "Yes, but you are in the same position as me. You can't know anything for certain, either."

Christian: "Well, not so fast... We have established that in order to know anything at all, we must either be omniscient, *or* have the revelation of someone who is. That is the Christian worldview! God is omniscient, and He has chosen to reveal some knowledge and truth to us through His Word, the Bible. I have a basis for knowing things for certain!"

Kindly point out to the non-Christian that he gave up knowledge, but that *you still have a basis for knowing things for certain!*

The Uniformity of Nature

Experiments with consistent results and successful predictions regarding our universe are only possible because of the *unceasing consistency* of our universe. Things like gravity, gas laws, and mathematical laws must remain constant for any experimental results to make sense! Only in the Christian worldview do we get this uniformity of nature:

- God made all things (Genesis 1:1, John 1:3)
- God upholds all things by His power (Hebrews 1:3)
- God is consistent (1 Samuel 15:29, Numbers 23:19)
- God is omnipresent (Psalm 139:7-8)
- God is beyond time and upholds the universe in a consistent fashion (2 Peter 3:8)
- God tells us that we can count on certain things to be true in the future (Genesis 8:22, Jeremiah 33:20-21)

Think about this: If the universe were really the chance product of a magical big bang, then why would it obey laws like those of mathematics and gravity? If everything was just the product of random chemical reactions, then we could not rely on mathematical laws, physics laws, chemistry laws, or anything else, to always be consistent! There would be no basis for things like gravity to be the same in the future as it is today. Scientists could not repeat experiments to establish theories according to the Scientific Method unless they could rely on the uniformity of nature. As is turns out, secular scientists must assume the Christian worldview while conducting tests and experiments because *they must rely upon the uniformity of nature to be true while conducting those very same experiments.* **Isn't it ironic that scientists who try to disprove Christianity though scientific experiments must first assume the Christian worldview?**

Absolute Morality

An absolute, universal moral code by which we have knowledge of right and wrong only makes sense if there is a sovereign God who has created rules for us, and to whom we are accountable. An absolute authority must be our source of morals. Why is this the case? Without an absolute source, morals are just relative. They would just be a convention that is arrived at arbitrarily by someone. Because of this, someone else could come up with a set of "morals" that would conflict with the first person. There would be no objective way to say who is absolutely right! We discuss in chapter 15 how to refute relative morality to show why God (the absolute authority) is necessary for absolute morality. This moral relativism is easy to refute (as we will learn later), but it is rampant in our post-modern society.

Reliability of Senses and Memory

The fact that the human mind is capable of rational thought and that our senses can reliably examine the universe makes sense- our omnipotent and omniscient God created the human mind and sensory organs! He is the reason why we can trust our senses!

> "So God created man in his own image, in the image of God created he him; male and female created he them" (Genesis 1:27)

"The hearing ear, and the seeing eye, the Lord hath made even both of them." (Proverbs 20:12)

"Then the Lord said to him, 'Who has made man's mouth? Who makes him mute, or deaf, or seeing, or blind? Is it not I, the Lord?" (Exodus 4:11)

"Who put wisdom in the heart, or gave the mind understanding?" (Job 38:36 HCSB)

Human beings coming about as a result of random chemical reactions, as in the evolutionary worldview, have no basis for sensory organs to work reliably and consistently.

Laws of Logic

What are laws of logic? They are "...fundamental laws upon which logic and rational thinking are based. The (following) three laws are thought to have originated with Aristotle, who believed that the laws are necessary conditions for rational thinking to occur. The three laws are the law of identity, law of non-contradiction, and law of the excluded middle."[4]

The first three laws of logic, as given and defined on CARM.org, are[3]:

1. The Law of Identity
    a. This law states that A is A. In other words, something is what it is. If something exists, it has a nature, an essence.

2. The Law of Non-Contradiction
    a. This law states that A cannot be both A and not A at the same time and in the same sense. In other words, something (a statement) cannot be both true and false at the same time and in the same way.

3. The Law of Excluded Middle
   a. This law states that a statement is either true or false. For example, my hair is brown. It is either true or false that my hair is brown.

Can you touch, see, or weigh the laws of logic? No! They are: *immaterial* (not made of matter), *universal* (apply everywhere), *abstract* (exist in thought without having a concrete existence), *and invariant* (do not change). These laws exist because of the nature of the Biblical God! Yet, all people, including unbelievers, presuppose these to work – but *they are just borrowing from the Christian worldview!* Think of the implications here- while logic is governed by laws, human reasoning can still be flawed. People don't always stay logical while using their reasoning abilities and giving arguments to make a case for what they are trying to prove or establish.

The Ultimate Proof of Christianity

In getting a basic understanding of presuppositional apologetics, if you have studied other religions and their gods, two things should become very apparent:

1. *Every religious worldview other than Christianity will fail one or both of the first two tests, "A" and "I", from the AIP test above, and*

2. *Every non-religious worldview will fail at least the "P" portion of the AIP test.*

This means that when we apply the AIP test to every religious worldview:

1. It cannot be arbitrary (without justification).
    a. The majority of religious systems have some element of being arbitrary, making it unsupported and false. They mostly rely on an arbitrary source, like a committee of men, to make decisions. For example, Catholicism has popes and bishops who change over time, which gives rise to thoughts and doctrines within Catholicism to change over time.

2. It cannot be inconsistent.
    a. Every religious system besides Christianity has contradictions within its doctrine, making it unreliable and false. For instance, Mormons used to believe that "black people" were cursed, but now the curse is gone.

3. It must satisfy the preconditions of intelligibility (such as having a basis for absolute knowledge, and absolute morality)
    a. Every non-religious worldview lacks an absolute authority (God), thus none of them can satisfy this requirement.

Of all the religious systems and non-religious systems that exist, only Biblical Christianity can satisfy the necessary requirements to be a valid worldview. It is no wonder why God tells us:

"The fear of the Lord is the beginning of knowledge: but fools despise wisdom and instruction." (Proverbs 1:7)

The ultimate proof of Christianity is this: one *cannot make sense of anything without the one, true, creator God (of the Bible) as our starting point!* The ability to have absolute knowledge comes from us starting with God! **One cannot reason to Him; He is the necessary starting point!**

Who Knows that the True God of the Bible Exists?

What does the Bible say about people who say that they don't believe God exists?

> "The fool hath said in his heart, 'There is no God.'" (Psalm 14:1, KJV)

> "A fool takes no pleasure in understanding, but only in expressing his opinion." (Proverbs 18:2)

Why does God say that it is foolish to say He doesn't exist? We have already established this regarding the preconditions of intelligibility in this chapter - one cannot have knowledge or even the ability to reason without God's existence. As Andrew Rappaport states, "An unbeliever must presuppose a Christian worldview (rely on God) to be able to deny Him." But, there is one more glaring reason why God calls unbelief foolish- the Bible is clear regarding people's knowledge of the existence of God.

> "For the wrath of God is revealed from heaven against all ungodliness and unrighteousness of men, who by their unrighteousness suppress the truth. For what can be known about God is plain to them, because God has shown it to them. For his invisible attributes, namely, his eternal power and divine nature, have been clearly perceived, ever since the creation of the world, in the things that have been made. So they are without excuse. For although they knew God, they did

not honor him as God or give thanks to him, but they became futile in their thinking, and their foolish hearts were darkened. Claiming to be wise, they became fools, and exchanged the glory of the immortal God for images resembling mortal man and birds and animals and creeping things." (Romans 1:18-23)

From this passage, it is clear that every person knows the true God that exists. Whether he claims God to be Buddha, Allah, or any other false god, or, if he claims that God doesn't exist, the Bible tells us that he does know the true God and he is suppressing the truth in unrighteousness (sin)! That person will be without excuse on Judgment Day. Furthermore, the people who say that they don't believe in God end up worshipping some type of false god/idol.

After reading and understanding this passage in Romans, it is easy to acknowledge the foolishness when people claim there is no God.

Battle of the Worldviews

According to the world, there are numerous worldviews out there, many of which overlap: evolutionism, naturalism, post-modernism, false religions and cults, and Biblical Christianity. We have addressed in this chapter how every one of them can be shown to be false, except Biblical Christianity. However, according to the Bible, how many worldviews actually exist?

"He who is not with me is against me, and he who does not gather with me scatters abroad." (Matthew 12:30)

"Because the carnal mind is enmity against God; for it is not subject to the law of God, nor indeed can be." (Romans 8:7)

"Adulterers and adulteresses! Do you not know that friendship with the world is enmity with God? Whoever therefore wants to be a friend of the world makes himself an enemy of God." (James 4:4)

Obviously, there are only 2 worldviews according to the Bible:

1. The Biblical God, and
2. Everything else (Unbelievers, Hindus, Buddhists, Roman Catholics, Jehovah's Witnesses, Mormons, etc.)

The Pretended Neutrality Fallacy

Non-Christians, whether they are agnostic or of a false religion, almost always want to meet you on "neutral ground." Their attempt is to get you to a spot of agreement with them, and then you can each argue from that supposed "neutral" spot. This, however, begs the question:

"Can we be neutral?"

What are you willing to give up to be neutral? Should you compromise on the Ten Commandments? Should you compromise on the existence of hell? What about the days of creation? Jesus's perfect, atoning sacrifice? The reality is, once we attempt to be neutral, we are on *their* turf! Think about Jesus's words here, written down by Matthew:

"He who is not with me is against me, and he who does not gather with me scatters abroad." (Matthew 12:30)

If you give up the Bible, then you think you are going to win the argument on their ground and by your own reasoning. Again, what does the Bible say?

- Romans 5:12-21- Our identity is either in Adam or in Christ (in peace or at enmity with Him)

- 1 Peter 2:11- Either in Flesh or in the Spirit

- Matthew 7:24-27 – House built on rock or sand (smart or foolish)

- John 14:6 – I am THE way, THE truth, THE life, no one comes to the Father but by me

No neutrality must apply in every way. As we have already established, the only way that we can even have absolute knowledge or the ability to correctly reason is when we start with God and the Christian worldview. Therefore, no neutrality must also apply in the topics of:

- Science, and the interpretation of it
- Apologetics
- Topic of Evolution
- Topic of Homosexuality
- Sanctity of human life

Dr. Jason Lisle[1] records two things that the late Greg Bahnsen, a highly regarded Christian apologist, said regarding neutrality: **"They're not (neutral), and you shouldn't be!**[1]**"** Always remember that advice while in a worldview discussion. The reality is that your opponent is not, nor ever will be, on neutral ground himself. As we learned earlier, he will have a worldview that he stands on to argue from. Yet, your opponent will always appeal to you to become neutral when you attempt to stand your ground. All he is trying to do is pull you onto his "home turf" and off of your Biblical worldview. Your opponent inherently knows (Psalm 14:1, Romans 1) that he cannot win the discussion otherwise.

By standing your ground on the Biblical worldview, you have the full armor of God, as we see in Ephesians 6:

"Finally, be strong in the Lord and in the strength of his might. Put on the whole armor of God, that you may be able to stand against the schemes of the devil. For we do not wrestle against flesh and blood, but against the rulers, against the authorities, against the cosmic powers over this present darkness, against the spiritual forces of evil in the heavenly places. Therefore take up the whole armor of God, that you may be able to withstand in the evil day, and having done all, to stand firm. Stand therefore, having fastened on the belt of truth, and having put on the breastplate of righteousness, and, as shoes for your feet, having put on the readiness given by the gospel of peace. In all circumstances take up the shield of faith, with which you can extinguish all the flaming darts of the evil one; and take the helmet of salvation, and the sword of the Spirit, which is the word of God." (Ephesians 6:10-17)

Five pieces of armor – the belt of truth, the breastplate of righteousness, shoes of readiness, the shield of faith, and the helmet of salvation – are all defensive pieces of armor while in a worldview discussion. **That protective armor is lost when accepting a "neutral ground." Most tragically, we would also lose our only offensive weapon- the Sword of the Spirit, which is the Word of God.**

Certainty about God

Understanding presuppositional apologetics should make one thing abundantly clear - the fact that w*e can know for certain God exists*. We need not "be fairly certain" or "be beyond a reasonable doubt." God's existence is the only way in which we can account for our ability to reason and be able to evaluate anything. Even "evaluating the evidences" presupposes Him. **The fact that you can read this book right now and understand it proves that He exists – with 100% certainty.**

Chapter Eight

Problems with Evolution

The Complexity of Organisms

According to evolution, organisms have been getting increasingly complex over billions of years, starting from a "simple" single-celled organism to the complex organisms that are seen today. *For this to have happened, millions-to-billions of mutations that increase functional genetic information would have been necessary.* These same mutations would also explain the speciation of different organisms (such as the development of all of the different breeds of dogs today). Is this what we actually observe in science? No! We see the exact opposite- mutations decrease information overall!

According to the Bible, God designed every organism. Whether it consists of a single-cell or it is of the intricacy of a human- the pinnacle and crowning glory of His creation- every organism is complex. **The variation that we see within a kind (such as different dogs within the canine kind) is mostly due to the genetic diversity that God built into the original kinds**. Not only is this Biblical, but this is also consistent with what we observe in science today.

The reason why we see all of the different characteristics in different breeds of dogs and the other members of the canine kind is NOT because mutations gave rise to new genetic information. We observe this genetic variation within the canine kind because the genes were already present in the original parent canine kind. The genes for short hair, medium length hair, and long hair were present from the beginning. The process of natural selection due to environmental factors is what eliminates certain characteristics from populations.

In the case of the canine kind, it follows that the long-haired canines in hot climates would be selected against, thereby eliminating it from the population of canines that moved to that area. Similarly, we would see the same thing happen in cold climates where the short-haired dogs would be selected against. It is no wonder why we see polar bears, known for thermal outer layers, adapted for cold climates, and other bears of the same kind with different adaptations in warmer climates. Thus, natural selection actually **decreases** the overall genetic information among populations. It does not increase it.

Is Evolution Scientific?

In Chapter 3, we learned about how scientific theories are developed using the scientific method:

1. **Observation-** One must make an observation of something
2. **Hypothesis / Prediction** – One must then construct a hypothesis about the observation
3. **Testable** – One must test the hypothesis to see if the observation is true
4. **Conclusion** – One must make a conclusion that the test confirms the original hypothesis
5. **Repeatable** – If the first test was successful, the test must be repeated multiple times to see if the results still lead to the same conclusion
6. **Theory** – Once the repeat tests continue to confirm the original hypothesis, the hypothesis becomes a theory.

We also learned that for evolution to be true, mutations that increase functional genetic information *must* occur. For organisms to evolve from pond scum to a single cell to the complex organisms that we see today, millions-to-billions of those mutations would have been needed. Let us attempt to apply evolution to the scientific method.

1. **Observation-** Scientists have observed organisms can adapt to their environments.
2. **Hypothesis / Prediction** – It is hypothesized that these minor changes can lead to macroevolution – where a

kind of organism can change into a different kind over time (such as dinosaurs into birds).
3. **Testable** – Testing is conducted to try and show that mutations which increase functional genetic information do occur.
4. **Conclusion** – As we learned earlier, while a very few of the observed mutations are beneficial, there is not a single mutation that has ever been shown to increase genetic information!
5. **Repeatable** – N/A
6. **Theory** – N/A

The scientific method stops in step 4 here. **The only possible mechanism for evolution to work, the ability for mutations to produce new functional genetic information, has never been observed!** Because of this, evolution shouldn't even be considered a theory! *Evolution cannot even be considered science!* (Let that statement sink in!)

Evolution is a Religion

How is this for some irony: We tell kids the story that a princess kisses a frog and the frog turns into a prince. We know that this is a fairy tale. Yet, some scientists try to tell us every day that evolution caused amphibians to turn into humans many years ago!

As Biblical creationists, we believe, "In the beginning…" With our well-reasoned faith (Hebrews 11), we know that the next word in that sentence is "God." We often get asked the sarcastic question, "Well, who made God?" It is postulated to try and invalidate our explanation for the origin of the universe. Yet, evolutionists also believe, "In the beginning…" They just believe that the next word is "something." They must also have faith – this is a religion! They believe that whatever was there on earth before life- whether it be dirt, pond scum, or something else- it turned into everything that we observe today.

Regarding the origins of the universe, they must also have faith. They believe that something had to be there in the beginning of the proposed "Big Bang." Something had to explode into everything – *it is impossible for nothing to burst into something!* Having said that, some scientists have postulated that nothing turned into something using quantum mechanics. But that doesn't make any scientific sense, either. Even in the supposed "nothing-turned-into-something" scenario, they must assume that scientific laws of quantum mechanics exist in the beginning, which makes it self-refuting. Even famed atheist, Lawrence Krauss, admits all of this (except the self-refuting part) in his book, *A Universe From Nothing.*

In all scenarios, something must have been eternally existent – God, matter, or the laws that caused particles to magically pop into existence. *They are **all** religions*; only ours has an eyewitness that told us how it happened!

The "Random Chemical Reactions" of Evolution

Physicist Lawrence Krauss, a world-renowned atheist, once said, "Every atom in your body came from a star that exploded. And, the atoms in your left hand probably came from a different star than your right hand. It really is the most poetic thing I know about physics: You are all stardust. You couldn't be here if stars hadn't exploded, because the elements - the carbon, nitrogen, oxygen, iron, all the things that matter for evolution and for life - weren't created at the beginning of time. They were created in the nuclear furnaces of stars, and the only way for them to get into your body is if those stars were kind enough to explode. So, forget Jesus. The stars died so that you could be here today."[1]

According to molecules-to-man evolutionism, we are just rearranged stardust, pond scum, and dirt. We are the result of many billions of random chemical reactions, from pond scum to humans. Evolutionists will accuse you of being dumb (among other things) if you don't believe in their "theory." There is an easy way to turn the conversation around. Consider the following dialogue that I have with many evolutionists on the street:

Christian: "So, you believe that you are the result of random chemical reactions, from pond scum all the way up to you?"

Non-Christian: "Yes"

Christian: "Then, it follows that you must be a bag of random chemical reactions."

Non-Christian: "Sure." (The person knows that, logically, he must agree with your statement.)

Christian: "Then, your thoughts must also be random chemical reactions, if you are just a bag of chemical reactions."

Non-Christian: "I guess so" (Again, the person has no choice but to agree with you.)

Christian: "If you are just a bag of random chemical reactions, then how do you have a basis for an absolute right and wrong? What about a basis for calling things true or false? You don't even have a basis for having absolute knowledge!'

You can now point out to the person that they gave up knowledge. Like the "knowledge of the universe" example in the last chapter, this is another way of showing that one can only know things for certain if they start with God.

On top of that, there is another consequence to these *naturalistic* thoughts. If a person is just a bunch of chemical reactions, then any thoughts from his brain are actually just obeying the naturalistic laws. People would merely be following the natural laws of physics, chemistry, and biology without any ability to give any real meaning to them. They certainly cannot make any moral claims regarding them!

Relative Morality and Evolution

As we established in chapter 7, an absolute, universal moral code by which we have knowledge of right and wrong only makes sense if there is a sovereign God who has created rules for us and to whom we are accountable. For morals to be absolute, they must be "projected down" onto us by an absolute authority- God.

With an evolutionary worldview, we are just rearranged pond scum through a series of random chemical reactions. Morals cannot be absolute. As Jay Lucas[2] points out so eloquently, any explanation that evolutionists give for the authority of morals falls short of an absolute standard. Thus, without God, all morals are just relative.

On top of that, in an evolutionary worldview, where we are just random chemical reactions, what is wrong with "dirt putting holes in pond scum?" In more common terms, why should murder be wrong? After all, this is just survival of the fittest, anyway. Abortion should be no problem! Homosexuality and homosexual marriage shouldn't matter. If God didn't create us male and female from the beginning, transgenderism should not be a concern. Even pornography, racism, and other moral issues all are fine w/o an absolute standard.

It is clear that evolutionism has catastrophic problems! If everything is the result of random chemical reactions, then there is nothing that can be absolutely defined in the first place. Therefore, nothing that happens in this world would have any meaning!

It must be stated here that things do have meaning. We are all made in the image of God (Genesis 1:27) and everyone has the absolute moral laws written on their heart (Romans 2:14-15). Evolutionists are just attempting to suppress what they know about God in their unrighteousness (Romans 1:18-22). Without the saving grace of Jesus Christ, they will stand without excuse (Romans 1:18-22).

Chapter Nine

The Reliability of the Bible

God is the fundamental starting point of Christianity. As Christians, He is our axiom, which, by definition, means that He is regarded as being established, accepted, and self-evidently true. One cannot prove this fact. Our thinking must start with Him, which necessarily includes His revealed Word, the Bible. While we cannot "prove" our axiom, this entire chapter is devoted to showing how our axiom can be accounted for.

The only other axiom possible is "not-God." Can this axiom be accounted for (justified)?

An Ultimate Standard (axiom)

What is an ultimate standard (axiom)? Where does it come from? How can it be proven? Consider these three necessities:

1. Everyone must have an ultimate (absolute) standard.
2. An ultimate standard cannot be proved by another standard. (Otherwise it becomes the ultimate standard itself).
3. An ultimate standard cannot be merely assumed.

Why are these three points necessary? First, **every thought that we have and every statement that we make assumes an ultimate standard**. The only way that we can *honestly* converse with someone is by assuming that we both *implicitly* agree on a set of principles (truths, morals, ideals, etc.) which govern the communications. And the only way our communications have the same meaning outside the conversation as within it is if that set of principles is *universal*! These principles, **in total**, make up the ultimate standard.

Second, an ultimate standard cannot be proved by something outside of the ultimate standard. Otherwise, the proof would become the ultimate standard. This doesn't make sense logically; if something "proves" the ultimate authority, then THAT is now the ultimate authority. This cycle of proofs never ends at an ultimate authority. This is demonstrated in the "infinite regress" model:

1. Start with a supposed ultimate standard. We will call that standard, "A".
2. In order to prove "A", we appeal to "B".
3. But, we are now stuck with needing to prove "B". Thus, we need "C" to prove "B".
4. Now, we need "D" to prove "C".
5. This leads to an "infinite regress" that doesn't end:
    - Need "E" to prove "D"
    - Need "F" to prove "E"
    - And so on ...

The problem here is clear: an infinite regress never ends! Because of this, none of the standards in the infinite regress is actually the ultimate standard. Only a true ultimate standard, an axiom, can end an infinite regress. We can justify God as our ultimate standard. As we will see throughout this chapter, an unbeliever cannot justify his "not-God" ultimate standard.

Third, an ultimate standard must have a basis because we have to be able to appeal to it objectively. If an ultimate standard would just be assumed, it presents catastrophic problems. An ultimate standard that is assumed by a person would be that *person's own presuppositions*, which he would need to be able to justify. However, this would inject "relativity" into his supposed ultimate standard. And once the standard does not come completely from an ultimate source, it becomes, by definition, a relative standard overall. **Any assumption leads to a relative standard, not an ultimate one.**

Why is relativity bad, anyway? Because, once relativity is injected into a supposed ultimate standard, we must question the basis for each person's presuppositions (where each of their standards originates). As an example, once two people have a disagreement on their standards, assuming they are mutually exclusive, the relativity is displayed, and there wouldn't be a basis for distinguishing which standard is actually the correct one. As we saw in chapter 7, the laws of logic shows that, at most, only one person could be right. In the case of God versus "not-God," only one can be right!

Because of the three necessities shown above, especially in light of the "infinite regress" model, the only way that we can demonstrate an ultimate standard is this: **an ultimate standard must prove itself!** It must be self-attesting! Many people will ask you to "prove" God. The problem with this challenge is that for God to be God, nothing can be higher. If there was something higher, or something that could "prove" God, then he is not the Christian God, who is omniscient and omnipotent, and is infinite in time, space, matter, knowledge, and power.

The ultimate standard is God -the one, true God of the Bible. How do we know this? He proves Himself - His Word is self-attesting! Consider the following verse about the ultimate standard:

> "For when God made a promise to Abraham, since he had no one greater by whom to swear, he swore by himself." (Hebrews 6:13)

It is clear that, as Voddie Baucham correctly states, "If (a person) would appeal to a higher authority to prove the Bible, then (he) would be *conceding* the fact that there is a higher authority."[1] **The Bible is, and must be, our highest objective authority because it is the Word directly from God, and He is the highest authority.**

The Self-Attesting Nature of the Bible

There are many great facts about the Bible that are consistent with its self-attesting nature. It was written by over 40 different authors, in 3 different languages, on 3 different continents, and over a period of 1500 years. Furthermore, these 66 books that make up the

Bible cover hundreds of topics by authors from different walks of life who, the majority of which never met one another. *What makes these facts amazing is that the books of the Bible are completely consistent with each other*! Complete unity!

In another eloquent statement by Voddie Baucham, he pronounces, "I choose to believe the Bible because it is a reliable collection of historical documents, written by eyewitnesses during the lifetime of other eyewitnesses. They report supernatural events that took place in fulfillment of specific prophecies, and they claim to be divine rather than human in origin."[1]

Historical Documents

While we already understand that the Bible is the only justifiable axiom, many people, especially scoffers, make the erroneous claim that the Bible can't be proven scientifically. While we have already addressed the issue in Chapter 3, it is worth mentioning again that historical accounts cannot be proven using science as the scientific method **requires** things to be tested, repeated, and verified. **History cannot be repeated to test and verify!** Instead, one must use the basic principles of historiography to determine the accuracy and reliability of the historical accounts contained in the Bible, as well as the Bible itself. These tests include the Bibliographical test, the Internal Evidence test, and the External Evidence test. Most of this chapter is devoted to the Internal Evidence test only because of the fact that the Bible must be an ultimate standard and, therefore, self-attesting. **This Internal Evidence test alone provides the proof that the Bible is completely true.** The Bibliographical and External Evidence tests merely provide evidence that confirms the historicity of the Bible. Some of these evidences are:

1. Over 25,000 archaeological digs have confirmed and affirmed the Bible[1],
2. The entire Bible was written down by the end of the 1st Century[1],
3. We have documents that date back to within a few decades of the original penning of the NT[1],

4. Using the writings (quotes and commentaries) of the early church fathers, we could reproduce the entire NT except for 11 verses[1],
5. The Dead Sea Scrolls, which are dated from around 250 B.C. to 68 A.D. and discovered between 1947 A.D. and 1956 A.D.[5], show that the OT that we have today is reliable and hasn't changed in the last 2000 years, as the "telephone game" scoffers try to claim happened. These Dead Sea Scrolls also prove that the book of Isaiah, which contains many prophecies about Jesus Christ, was actually written by the prophet Isaiah around seven hundred years before the events happened. Because the prophecies contained in Isaiah are perfectly accurate, scoffers had tried to claim that the book of Isaiah was written after the events occurred.

The claim that the Bible was copied many times and changed over time, like what happens in the "telephone game," is something that needs to be addressed here. The "telephone game" is where information is passed on audibly, without being written down or checked in each passing of the information to the next person. Bible replication did not occur this way. Each new copy of the Bible was written down, not passed audibly, and was checked against the original.

In regards to new translations of the Bible over the years, the translators did not go to the previous translation to make the new one. In each new translation, the translators go back to all of the oldest source documents.

The copying of the Bible and the making of new translations can best be described using a cookie recipe that gets shared among many people. The copying and/or translation of the Bible was not a person getting the recipe from Grandma Smith, then orally passing it to another person years later, who then passes it one to another person years later, and so on, where the recipe can get changed over time to where everyone has a slightly different recipe. Instead the copies of the Bible are the result of each person going to the source of the recipe, Grandma Smith. Every good Bible in existence today is the result of going to the oldest source documents in order to translate it into the current language and then copied exactly to replicate it.

Reliable Collection of Historical Documents that are divine in origin

It is of utmost importance that we realize the Bible was not composed of just one isolated author that claims he received special instruction from God (which, by the way, seems to be the recipe for most cults and false religions, such as Mormonism, Islam, and Jehovah's Witnesses). The Bible is a **collection** of books written by many authors who documented real, historical events. Moreover, the origin of the words, sentences, and thoughts is of God and not the human authors themselves.

The false religions, most of which have one initial author, still have internal contradictions. Yet, over forty people, of different walks of life and education, wrote the sixty-six books of the Bible over 1500 years with no real contradictions! Many websites claim to catalog different contradictions in the Bible, but all of them have been answered and most of these seeming contradictions are misinterpretations on the part of the reader.

> "And we have the prophetic word more fully confirmed, to which you will do well to pay attention as to a lamp shining in a dark place, until the day dawns and the morning star rises in your hearts, knowing this first of all, that no prophecy of Scripture comes from someone's own interpretation. For no prophecy was ever produced by the will of man, but men spoke from God as they were carried along by the Holy Spirit." (2 Peter 1:19-21)

Peter is making it very clear in this passage that ALL Scripture is divine in nature and not derived by humans. Not a single author of the Bible makes a claim that any part of Scripture is of human nature alone. But there is more to this point – the Bible is overflowing with phrases that say the Bible is the written Word of God. When doing a phrase search in the Bible regarding this, it is seen that:

> "Some thirty-eight hundred times the Bible declares, 'God said,' or 'Thus says the Lord' (e.g. Ex. 14:1; 20:1; Lev. 4:1; Num. 4:1; Deut. 4:2; 32:48; Isa. 1:10, 24; Jer. 1:11; Ezek. 1:3; etc.)."[4]

It is obvious that the Bible makes the unmistakable claim that it is the written Word of God!

How do we know that the Bible is not changed, textually, from the earliest source documents? Does Bart Ehrman have a leg to stand on when he claims that there are over 400,000 variants between the oldest source documents, making the Bible unreliable? We will tackle the issue of textual criticism in the next chapter, by Pastor Andrew Rappaport.

This entire chapter, and especially this section, is devoted to confirming the complete truth of the Bible and that it is the Word of God written down in perfect precision by men. They recorded the events of the Bible *exactly* as God wanted it recorded. Yet, oftentimes while witnessing, you will hear the claim from an unbeliever that we can't believe the Bible because mere men wrote it. They are making the claim that imperfect man wrote the Bible and it is not divine; therefore, it cannot be trusted. The issue is not really about the Bible being written by men. This is solely an issue of trustworthiness.

While we as Christians can easily reject that claim, it is imperative to point out the fallacy of the unbeliever's statement. You must merely tell him that, "By that line of reasoning, you cannot know anything at all because everything you have ever learned has been through the books written by mere man and the observations of mere man, including your own observations." It successfully closes the mouth of the unbeliever. Again, the Bible doesn't claim to be written by man – it is merely written down or recorded by man.

Written by Eyewitnesses During the Lifetime of other Eyewitnesses and Divine in Origin

The Bible is also unique because the authors were either the actual eyewitnesses of the historical events that they wrote about, or they received the information from the actual eyewitnesses of the events.

> "For we did not follow cleverly devised myths when we made known to you the power and coming of our Lord Jesus Christ, but we were eyewitnesses of his majesty. For when he received honor and glory from God the Father, and the voice was borne to him by the Majestic Glory, 'This is my beloved Son, with whom I am well pleased,' we ourselves heard this very voice borne from heaven, for we were with him on the holy mountain." (2 Peter 1:16-18)

Peter makes the claim in this passage that he as well as others saw all of the events that occurred regarding Jesus' life, miracles, death, resurrection, and ascension, personally. He even verifies the historical account of the Mount of Transfiguration and makes the claim that the Father spoke to him, James, and John! (Matthew 17:1-8, Mark 9:2-8, and Luke 9:28-36) Remember also that this passage (2 Peter 1:16-18) immediately precedes the passage above regarding the fact that ALL Scripture is divine in nature and not derived by humans (2 Peter 1:19-21).

The Gospel of Luke is also of interest in this discussion, as noted by an article titled "Doctor Luke" written by Henry Morris:

> "Luke—the author of the third Gospel and the book of Acts—is of special interest for several reasons. He was the only Gentile who wrote any of the books of the Bible. Furthermore, he was the only scientist among the writers. He is also recognized as a great historian, with his excellent accounts of the key events of the most important era in the history of the world. He also was undoubtedly a devoted Christian, a truth

especially demonstrated by his unselfish service and companionship to the apostle Paul. Finally, he was probably the first Christian apologist, zealously concerned to
defend and establish the absolute truth of the gospel of Christ."[2]

Luke, who didn't personally witness the events that he wrote about, was a careful historian who collected and wrote down the historical accounts of many eyewitnesses of the actual events. His Gospel starts off this way:

"Inasmuch as many have undertaken to compile a narrative of the things that have been accomplished among us, just as those who from the beginning were eyewitnesses and ministers of the word have delivered them to us, it seemed good to me also, having followed all things closely for some time past, to write an orderly account for you, most excellent Theophilus, that you may have certainty concerning the things you have been taught." (Luke 1:1-4)

So, Luke isn't just making a claim that he was the eyewitness of what he wrote, but that this was the compilation of multiple eyewitnesses of these historical events! As a doctor who obviously wrote in great detail, he had no problem in believing the completely congruent testimonies of the eyewitnesses that he collected his info from!

Apostle John also starts off one of his books with eyewitness claims:

"That which was from the beginning, which we have heard, which we have seen with our eyes, which we looked upon and have touched with our hands, concerning the word of life—the life was made manifest, and we have seen it, and testify to it and proclaim to you the eternal life, which was with the Father and was made manifest to us— that which we have seen and heard we proclaim also to you, so that you too may have fellowship with us; and indeed our fellowship is with the Father and with his Son Jesus Christ." (1 John 1-3)

What is also remarkable about this is how John's previous book, The Gospel of John, comes to a conclusion. In the second last chapter, he states:

"Now Jesus did many other signs in the presence of the disciples, which are not written in this book; but these are written so that you may believe that Jesus is the Christ, the Son of God, and that by believing you may have life in his name." (John 20:30-31)

John makes the claim that of all the signs and miracles of Jesus that he wrote of, there are many others not written down, but that these **eyewitnessed, historical events** were written so that people would believe in Jesus Christ and have life in Him. Then, John ups the ante while concluding his Gospel:

"This is the disciple who is bearing witness about these things, and who has written these things, and we know that his testimony is true. Now there are also many other things that Jesus did. Were every one of them to be written, I suppose that the world itself could not contain the books that would be written." (John 21: 24-25)

John wrote this down after just finishing his historical account of Christ revealing himself again to the disciples by the Sea of Tiberias, the third time that Jesus was revealed to the disciples after he was raised from the dead.

It is clear that the Bible was written down by either the eyewitnesses of the historical accounts or by people who reported the events observed by the actual eyewitnesses. But that is not all! **These**

historical accounts of eyewitnesses were written down *during the lifetime* of other eyewitnesses. Why is this so important? It is because the writings of the Bible's authors were falsifiable. Other people living at the time could have disputed and falsified the written testimonies of the Bible's authors! Consider the following passage:

> "For I delivered to you as of first importance what I also received: that Christ died for our sins in accordance with the Scriptures, that he was buried, that he was raised on the third day in accordance with the Scriptures, and that he appeared to Cephas, then to the twelve. ⁶Then he appeared to more than five hundred brothers at one time, most of whom are still alive, though some have fallen asleep. Then he appeared to James, then to all the apostles. Last of all, as to one untimely born, he appeared also to me. For I am the least of the apostles, unworthy to be called an apostle, because I persecuted the church of God. Whether then it was I or they, so we preach and so you believed. (1 Corinthians 15:3-9, 11)

In verse 6, Paul appeals to the fact that Jesus appeared to more than 500 people after the Resurrection, most of whom are still alive. **Paul was giving testable and certifiable proof of the Gospel, and specifically, the Resurrection – go and ask the other hundreds of *still alive* eyewitnesses!**

Supernatural Events That Took Place in Fulfillment of Specific Prophecies

The Bible is full of supernatural events that are reported by the authors. Some of the prominent events include:

1. Moses crossing the Red Sea,
2. The Transfiguration,
3. The many claims that Jesus healed the sick,
4. That Jesus raised people from the dead, and
5. Jesus's own death and resurrection.

The best facet of the Bible's self-attesting nature is regarding the prophecy of Jesus Christ, including those involving some of the supernatural events of the Bible. According to Josh McDowell, author of *More Than a Carpenter,* there are sixty major messianic prophecies contained in the Old Testament, as well as 270 ramifications thereof. Most Christians are aware of the most popular ones, including the genealogies contained in the OT and Isaiah 53, which was written around 700 years before the birth of Christ and several centuries before any ever saw a crucifixion, as it wasn't even invented yet! The chances of just eight of the prophecies to be fulfilled in one person (let alone them all), is just 1 in 100,000,000,000,000,000! Yet, Jesus Christ fulfilled them all!

On multiple occasions in the Gospels, Jesus says that the OT points to Him.

> "But all this has taken place that the Scriptures of the prophets might be fulfilled." (Matthew 26:56)

> And he said to them, "O foolish ones, and slow of heart to believe all that the prophets have spoken! Was it not necessary that the Christ should suffer these things and enter into his glory?" And beginning with Moses and all the Prophets, he interpreted to them in all the Scriptures the things concerning himself. (Luke 24:25-27)

> Then he said to them, "These are my words that I spoke to you while I was still with you, that everything written about me in the Law of Moses and the Prophets and the Psalms must be fulfilled." Then he opened their minds to understand the Scriptures, and said to them, "Thus it is written, that the Christ should suffer and on the third day rise from the dead, and that repentance and forgiveness of sins should be proclaimed
> in his name to all nations, beginning from Jerusalem." (Luke 24:44-47)

> "And the Father who sent me has himself borne witness about me. His voice you have never heard, his form you have never

seen, and you do not have his word abiding in you, for you do not believe the one whom he has sent. You search the Scriptures because you think that in them you have eternal life; and it is they that bear witness about me, yet you refuse to come to me that you may have life. I do not receive glory from people. But I know that you do not have the love of God within you. I have come in my Father's name, and you do not receive me. If another comes in his own name, you will receive him. How can you believe, when you receive glory from one another and do not seek the glory that comes from the only God? Do not think that I will accuse you to the Father. There is one who accuses you: Moses, on whom you have set your hope. For if you believed Moses, you would believe me; for he wrote of me. But if you do not believe his writings, how will you believe my words?" (John 5:37-47)

Let us consider just a few of the prophecies from the Old Testament, all of which were fulfilled in the NT by Jesus Christ. It is important to note that, for the sake of argument, a person intending to do so during his lifetime could have purposely fulfilled some of these prophecies. However, many of the prophecies are ones that the person had no control over, such as the place of birth. Remember, every one of these prophecies was written at least 400 years before Christ being born on Earth.

1. **Garments Parted and Lots Cast**

Prophecy: "They divide my garments among them, and for my clothing they cast lots." (Psalm 22:18)

Fulfillment: When the soldiers had crucified Jesus, they took his garments and divided them into four parts, one part for each soldier; also his tunic. But the tunic was seamless, woven in one piece from top to bottom, so they said to one another, "Let us not tear it, but cast lots for it to see whose it shall be." This was to fulfill the Scripture which says, "They divided my garments among them, and for my clothing they cast lots." So the soldiers did these things. (John 19:23-24)

2. His Side Pierced

Prophecy: "And I will pour out on the house of David and the inhabitants of Jerusalem a spirit of grace and pleas for mercy, so that, when they look on me, on him whom they have pierced, they shall mourn for him, as one mourns for an only child, and weep bitterly over him, as one weeps over a firstborn." (Zechariah 12:10)

Fulfillment: "But one of the soldiers pierced his side with a spear, and at once there came out blood and water. He who saw it has borne witness—his testimony is true, and he knows that he is telling the truth—that you also may believe. For these things took place that the Scripture might be fulfilled: "Not one of his bones will be broken." And again another Scripture says, "They will look on him whom they have pierced." (John 19:34-37)

3. Darkness Over the Land

Prophecy: "And on that day," declares the Lord God, "I will make the sun go down at noon and darken the earth in broad daylight." (Amos 8:9)

Fulfillment: Now from the sixth hour there was darkness over all the land until the ninth hour. (Matthew 27:45)

4. Sold for Thirty Pieces of Silver

Prophecy: Then I said to them, "If it seems good to you, give me my wages; but if not, keep them." And they weighed out as my wages thirty pieces of silver. (Zechariah 11:12)

Fulfillment: Then one of the twelve, whose name was Judas Iscariot, went to the chief priests and said, "What will you give me if I deliver him over to you?" And they paid him thirty pieces of silver. (Matthew 26:14-15)

5. The Savior to be born in Bethlehem

Prophecy: But you, O Bethlehem Ephrathah, who are too little to be among the clans of Judah, from you shall come forth for me one who is to be ruler in Israel, whose coming forth is from of old, from ancient days. (Micah 5:2)

Fulfillment: And Joseph also went up from Galilee, from the town of Nazareth, to Judea, to the city of David, which is called Bethlehem, because he was of the house and lineage of David, to be registered with Mary, his betrothed, who was with child. And while they were there, the time came for her to give birth. And she gave birth to her firstborn son and wrapped him in swaddling cloths and laid him in a manger, because there was no place for them in the inn. (Luke 2:4-7)

6. The Savior was Born of a Virgin

Prophecy: "I will put enmity between you and the woman, and between your offspring and her offspring; he shall bruise your head, and you shall bruise his heel. (Genesis 3:15)
(The reference to "her offspring" is due to the fact there is no earthly father- otherwise, the offspring would be in reference to the father)

Prophecy: Therefore the Lord himself will give you a sign. Behold, the virgin shall conceive and bear a son, and shall call his name Immanuel. (Isaiah 7:14)

Fulfillment: All this took place to fulfill what the Lord had spoken by the prophet: "Behold, the virgin shall conceive and bear a son, and they shall call his name Immanuel." (Matthew 1:22-23)

Fulfillment: In the sixth month the angel Gabriel was sent from God to a city of Galilee named Nazareth, to a virgin betrothed to a man whose name was Joseph, of the house of David. And the virgin's name was Mary. And he came to her and said, "Greetings, O favored one, the Lord is with you!" But she was greatly troubled at the saying, and tried to discern what sort of greeting this might be. And the angel said to her, "Do not be afraid, Mary, for you have found favor with God. And behold, you will conceive in your womb and bear a son, and you shall call his name Jesus. He will be great and will be called the Son of the Most High. And the Lord God will give to him the throne of his father David, and he will reign over the house of Jacob forever, and of his kingdom there will be no end." And Mary said to the angel, "How will this be, since I am a virgin?" And the angel answered her, "The Holy Spirit will come upon you, and the power of the Most High will overshadow you; therefore the child to be born will be called holy—the Son of God. (Luke 1:26-35)

The Bible also contains a number of prophecies that marked the time period in which the Messiah must come – while the Temple of Jerusalem was still standing. This is shown in several OT verses: Malachi 3:1, Psalm 118:26, Daniel 9:26, Zechariah 11:13, and Haggai 2:7-9. The Temple was destroyed in 70 AD, and has not since rebuilt![3]

The last set of prophecies of great interest reference the royal lineage of Jesus Christ, which a normal person would have no ability to fulfill in and of himself.

1. Born of the seed of a woman (Genesis 3:15)
2. Of the line of Shem (Not Ham or Japheth) when the world was repopulated by only these three lineages after the Flood (Genesis 9,10)
3. Of the line of Abraham (Genesis 12:1-3, 17:1-8, 22:15-18)
4. Of the line of Abraham's second son, Isaac (Genesis 17:19-21, 21:12)
5. Of the line of Jacob (Genesis 28:1-4, 35:10-12, Numbers 24:17)
6. Of the Tribe of Judah, which was one of twelve total tribes (Hebrews 7:13-14, Matthew 1:1-16, Isaiah 11:1, Genesis 49:10)
7. Of the line of Jesse (Isaiah 11:1-5)

8. Of the house of David, who is one of 8 sons of Jesse (Matthew 1, Luke 3, 2 Samuel 7:11-16)

"When your days are fulfilled to walk with your fathers, I will raise up your offspring after you, one of your own sons, and I will establish his kingdom. He shall build a house for me, and I will establish his throne forever. I will be to him a father, and he shall be to me a son. I will not take my steadfast love for him, as I took it from him who was before you, but I will confirm him in my house and in my kingdom forever, and his throne shall be established forever." (1 Chronicles 17: 11-14)

This is just a sampling of the over sixty major messianic prophecies contained in the Old Testament, as well as 270 ramifications thereof. The most awesome part is:

Jesus Christ fulfilled them all, **as we find in the New Testament!**

Chapter Ten

Biblical Reliability of the Text

The following chapter is adapted with permission from Pastor Andrew Rappaport's book: *What Do We Believe?*

Biblical Reliability

Many people attack Christianity due to a lack of an understanding about how we got the Bible. Many try to claim we cannot trust the Bible, our only authority from God. The only way we can know about God objectively and absolutely is if He reveals Himself to us in some form of universal communication. The presupposition of Christianity is that God exists, He has spoken, and He has given us a reliable book, the Bible.

This chapter will cover a topic that most people with a seminary education never deal with unless challenged by an unbeliever. This can often be a very technical study. However, this chapter will simplify the complex as much as possible and provide the reader with the proper information to adequately know that the Christian Scriptures are not only reliable but are in fact the most reliable documents of ancient history.

Refuting the Critics

One of the most common arguments non-believers make is that men wrote the Bible. It is a trick question because men did write the Bible, but God also wrote it. Non-believers claim that if men wrote it, the Bible cannot be trusted. However, men have written everything taught to us, and they trust other things they have learned from men. What makes the Bible trustworthy is that God, who cannot lie and is faithful and true, wrote it. We trust in an infallible Author.

Christians get their authority from the Word of God alone, Sola Scriptura. The Bible is without error and without flaw in the original writings. However, the problem is we do not have any original manuscripts of the Bible presently, at least none we know of. The many manuscripts we have today do have some variants between them.

Daniel Wallace states, "A textual variant is simply any difference from a standard text (e.g., a printed text, a particular manuscript, etc.) that involves spelling, word order, omission, addition, substitution, or a total rewrite of the text." Each time we have a variant between manuscripts, it becomes what is called a "variant reading". We can have several variant readings for the same word between many manuscripts. You could have some manuscripts that read the same Greek phrase as, "Our Lord Jesus Christ", "Jesus Christ our Lord", "the Lord Jesus," or "our Lord Christ," and these would count as four variant readings.

The more manuscripts you have, the more possible variant readings you could have. If you only have one manuscript, then you have no variants, but you cannot tell if that one is a copy or if it is the original because you have nothing to compare. Therefore, the larger the number of manuscripts we have, the better is the opportunity to discover where all these variants could be in the text.

Scholars argue that the number of variant readings is about 400,000 in the New Testament. That sounds like a large number, especially when you consider that there are only about 138,200 words in the Greek New Testament. This would mean there is three times the number of variant readings than there are words in the New Testament.

The reason for this is that some people count each change, even of the same word between multiple manuscripts, as an individual variant. Therefore, a single word could have dozens of variants. It would be more accurate to only count the number of words that actually changed between all the different manuscripts.

So, we need a more precise way to define things - we need a base text to compare against all the other manuscripts. Fortunately, Hodges and Farstad have provided that for us in: *The Greek New*

Testament according to the Majority Text. They listed in the footnotes all the places where the majority of manuscripts disagreed with the Nestle-Aland text. When you compare the words with a base text, the total comes to 6577.

The fact that there are different manuscripts or variants between manuscripts should not cause concern regarding the authority of the Scripture. As we will see, due to the large number of manuscripts, we can identify where these changes occurred. This knowledge reveals that not a single Biblical doctrine is affected by these changes. Furthermore, with most of these changes, we can easily determine the original meaning of the text.

If you read Bart Ehrman, he makes it sound like the Bible's original meaning is lost forever. In his book, *Misquoting Jesus*, he tries to make this point. Ehrman's greatest example why we cannot determine the original meaning of the Bible is that some manuscripts refer to Jesus as a "carpenter" and others as the "son of a carpenter." In this example, the meaning of the text does change, and we cannot get back to the original meaning. But, there is no biblical doctrine based on whether Jesus is a carpenter or the son of a carpenter. In fact, both could be true. If this is the best example of a lost "original meaning," then we have nothing to fear.

In the book, *Zealot: The Life and Times of Jesus of Nazareth*, Reza Aslan tries to make the same case, and yet you can throw out all his work prior to chapter one. In the introduction to his book, he states that he is basing his research on a document called Q (German for Qulle or "source"). Then he states, "Although we no longer have *any* physical copies of this document, we can infer its contents by compiling those verses that Matthew and Luke share in common but that do not appear in Mark." In other words, he has no evidence for everything he is stating. The reality is that there are no physical copies of Q, and there are no references to this Q in any ancient text or history. There is no evidence that this document, Q, ever existed. Scholars make up this document Q and then criticize the Bible based on a document that has no record of ever having existed. Most people would call that a fairytale, a grasping at straws.

Category of Variants

As we talk about these variants, it would be helpful for us to define the types of variants and the problems or lack thereof that they cause (see figure 1). The majority of these variants, 75%, are spelling errors or punctuation. One thing to remember is that there was neither punctuation nor spacing between words in the manuscripts for the first 800 years of the New Testament. The misspelling of words can change the meaning of a text base. Also, the adding of spaces can change the meaning of the text, depending on where you add the spacing. As an example, think about the letters, "GODISNOWHERE." You can add spaces to make two very different sentences: "God is now here," and "God is nowhere". Which is right? The context is often used to determine this. So with these spelling errors, we can easily determine the original spelling in most cases.

The second largest category of variant, 19%, is not meaningful. This means that the variant does not change the meaning of the text in any way. Therefore, these do not represent an issue for the understanding of the meaning of the Scriptures. The third category, 5%, is those variants that are meaningful but not viable. Viable means that we cannot get back to the original text. Therefore, with these variants, we can get back to the original text either by the context or more often from the numbers of other manuscripts.

The last type of variant is the smallest, less than 1%, and most significant. These variants are meaningful and viable. These are the only ones that present a problem, because they do affect the meaning of the text, and we cannot get back to the text. These are the only variants that cause any problems for Biblical scholars and those that do textual criticism.

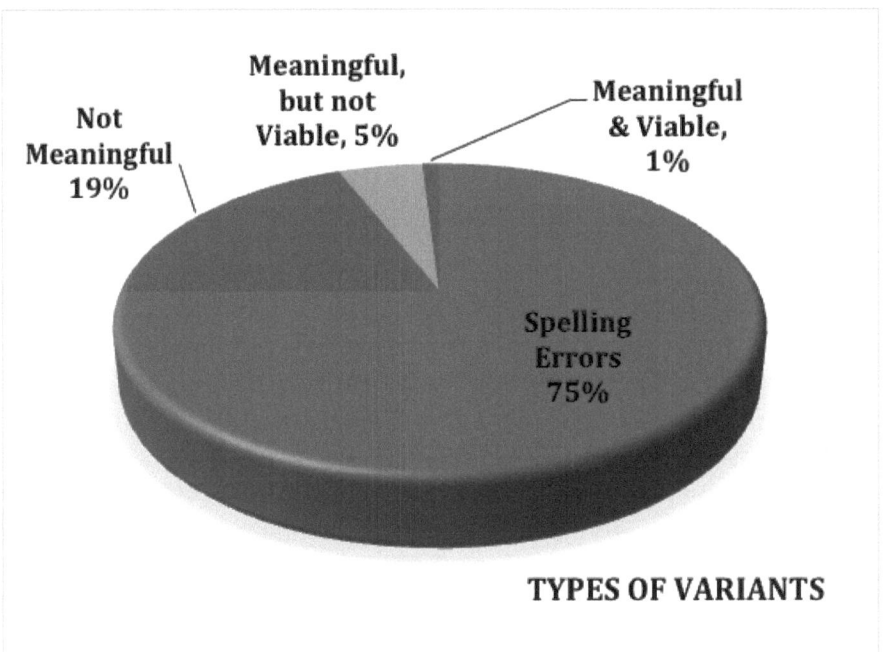

Figure 1

Out of all the variants, 99% of variants either do not impact the meaning of the text, or we can get back to the original meaning. Therefore, only 1% of the variants need to concern us. That is 1% of the 6577 textual variants (about 66 meaningful textual variants) out of 138,162 words, which is about 0.0476% of the New Testament. Now, all of a sudden, when put in context, that number of 400,000 variants does not seem so big. Therefore, we can say that the New Testament is 99.95% accurate.

Dispelling the Myth

We must also dispel a myth that many of these liberal scholars and professing atheist bloggers tell. The Bible translation was not like the telephone game, where one person makes one copy with a few mistakes and passes it along to another person who would make another copy but with some more mistakes and so on until the final product is nothing like the original. This is how many depict that the Bible was being disseminated and transmitted to us. However, that is not accurate to the means of dissemination and transmission of the Scriptures.

In the Old Testament, Jewish scribes had a process of copying that was painstaking in detail. The Jewish scribes would copy the Bible; they would take every single letter and count it, then every single word and count it. So with every letter, they would mark off on a chart to increase the number of times that letter was used. They would do the same with the words. Therefore, at the end of the copy, they could tell the frequency of every letter and every word. Then they could compare that to a chart that recorded all the correct number of letters and words to know if they got anything wrong. If they had more than three mistakes, then that copy could not be used for the synagogue readings. If it had even one mistake, it could not be used for copying.

The New Testament was different. Remember, this was a time before the printing press, copy machines and (believe it or not) the Internet. They had to copy each one by hand. The first century Christians wanted to get the good news of the Gospel out to as many people as possible, and as quickly as possible. So they were not as concerned with accuracy, but they were concerned with speed and distribution.

They did not play the telephone game, where you have only one person handling the message at a time and passing on a corrupted version of what they received. The copying of the New Testament was very different. Someone like Paul would write a letter of the New Testament and would have several people making copies at the same time. They would make multiple copies and give it to others to make multiple copies. These copies were spread through the region so that they could be copied and spread around. Therefore, when there were changes, we could look at the geographical location to see how this change was fostered.

Matt Slick provides a great illustration of this process, and he shows how we can determine what and where a change occurred from his website, CARM.org (see figure 2). When a word would go missing from a copy, we can compare that manuscript with the others copied in that geographical area. We then could see that somewhere in that location there was a missing word in the manuscripts, but not elsewhere around the world. As we gather thousands of manuscripts from around the world, we can compare these local changes to others and discover the change. These groups of local manuscripts have become known as manuscript families.

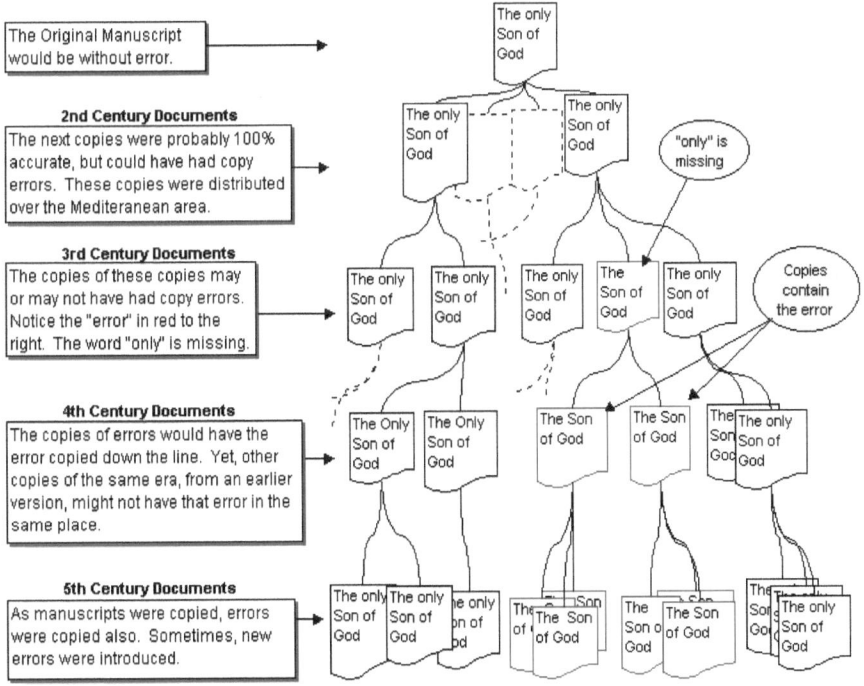

Figure 2

Factors for reliability

There are three factors that weigh heavy in the science of textual criticism:

1. Closeness of the copy to the original writing,
2. The number of manuscripts, and
3. The geographical location.

We can now see how the number of copies and the geographical location of the copies play a role in determining the original text. The age of the copy is important because the closer it is to the original, the less likely that it contains changes. Remember that changes occur over time: the less time that you have, the fewer changes you should have.

The age of the manuscript is very helpful to assume that there would be fewer changes. Also, there are times when we can rely on the early church fathers for help. In some early manuscripts, the number of man mentioned in the book of Revelation was 616, instead of 666. Now, that does not affect any doctrine, except for maybe a large genre of Christian end-times movies and the "Left Behind" book series. However, we have a second-century church father, Irenaeus, who spoke of this early textual variant. Now, by Irenaeus's time, he actually addresses this variant, and he said the better manuscripts seemed to point to 666. So the early church fathers help us to identify the corrections to some early variants.

We are consistently discovering more and more Greek manuscripts, and the numbers are adding up with many of them being very close to the original writing. With modern technology, we can now read fragments of manuscripts from papyrus that was washed and reused. This was a common practice because the materials used to write on were very expensive. Therefore, the oldest manuscripts often were reused, as good environmentalists would do. But, in this case, they were more like capitalists who wanted to save some money. Another location for the more ancient manuscript fragments is mummies' tombs. Older Greek manuscripts would be used as scrap paper to mummify people. If people wanted to edit the text of

Scripture, the trash is a place that would not be affected and a good place to look.

We have discovered a fragment of the book of John from 125 AD (called P52). This is within 30+ years of its writing. This is extremely close to the original, as we will see when compared to other ancient manuscripts. Some more recent discoveries have given us some fragments within the first century, only a few generations from the original writing.

We have in the order of 5,700+ catalogued Greek manuscripts, the average of which is about 200 pages. We have over 125 Greek manuscript witnesses within the first 300 years after its writing. As many uneducated critics argue that the Bible was edited in the fourth century, these manuscripts become important to prove these critics incorrect. Also, many Muslims argue that the Bible was corrupted after the time of Jesus. However, the Quran stated that the Bible can be trusted (Q 5:44-48), and that was taught during the life of Muhammad (610-632 AD). If this corruption actually occurred, we would be able to know that from these much earlier manuscripts.

"It is not just the Greek [manuscripts] which count, either. Early on, the [New Testament] was translated into a variety of languages—Latin, Coptic, Syriac, Georgian, Gothic, Ethiopic, Armenian. These translations of the New Testament can help to get back to the original meaning of the text sometimes. There are more than 10,000 Latin [manuscripts] alone."[1] When you examine all of the manuscripts, the best guess is that we have over 20,000 manuscripts of the New Testament. However, if you did not use any of these manuscripts, you could recreate the whole of the New Testament, except for 11 verses, from the quotations of the early church fathers alone, during 150-200 AD. There are over 1 million quotations from the early church fathers. When you add all the manuscripts, fragments and quotations from the New Testament, you approach almost 70,000 copies. Suddenly, that is a very large number of New Testament references.

With all this information, it should be clear at this point that the common argument that the Bible was later edited and disseminated around the world in 300-500 A.D. is false. It is interesting that we trust other ancient documents to tell things like the history of Julius Cesar or the wars of the Greeks. How does the New Testament compare to other ancient and yet trusted documents?

We can examine the New Testament compared to Homer, Demosthenes, Herodotus, Plato, Tacitus, Caesar and Pliny (see figure 3). It is obvious that the New Testament is way more reliable than any of its closest neighbors of ancient documents, whether it is by the number of manuscripts or the time gap between the original writing and the nearest manuscript in time.

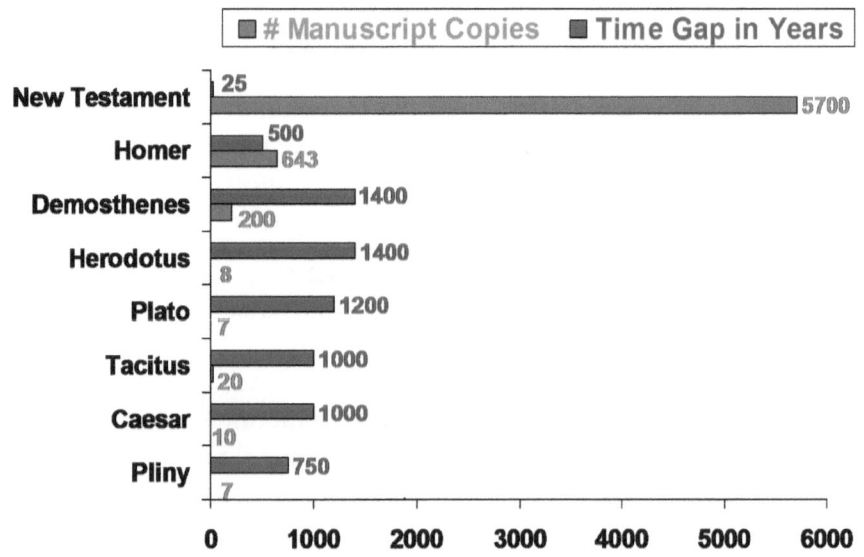

Figure 3

Mark Barry has a great visualization that brings much of this discussion together in one picture. It helps to see how the New Testament overshadows all of the other ancient documents (see figure 4). In this, he uses the size of the yellow dot to represent the number of manuscripts and the closeness to the center, and the black dot to represent the closeness of the writing to its earliest existed copies.

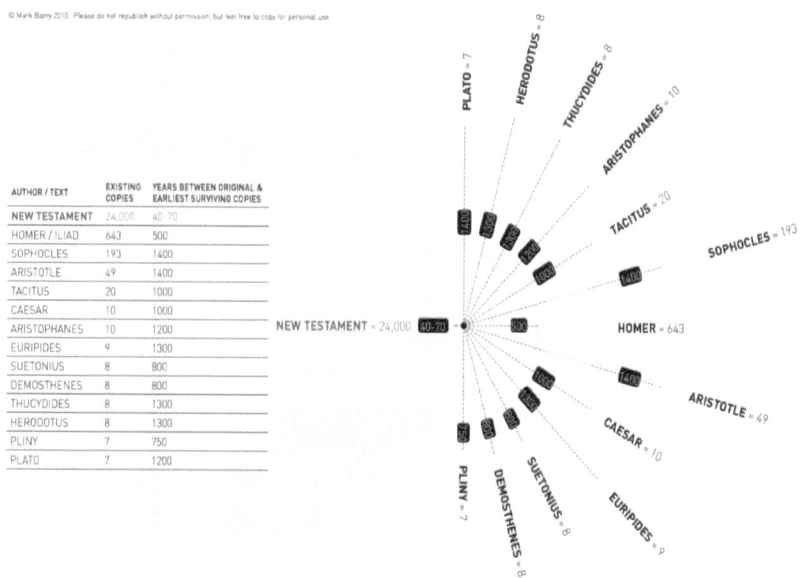

Figure 4

With all this information, we can rest on the fact that very little of the New Testament is in question, and even less of the Old Testament is. Due to the very large number of manuscripts, it helps us to identify every possible variant location in the Bible, and with that to know that not one of them affects any major Christian doctrine. Knowing that we can trust in the reliability of the Bible is different than understanding its meaning. That is the study of hermeneutics and is beyond the study of this book. However, a good resource for learning how to learn the principles of hermeneutics, to rightly interpret God's Word, can be found in the 20-lesson class on the Striving for Eternity Academy, courtesy of the ministry's website. (www.StrivingForEternityAcademy.org)

Chapter Eleven

How the Gospel is Affected by the Evolutionary Argument

The Spread of Evolution

Darwin was not the first evolutionist.

"Aristotle classified all living organisms hierarchically in his great scala naturae or Great Chain of Being, with plants at the bottom, moving through lesser animals, and on to humans at the pinnacle of creation, each becoming progressively more perfect in form."[1]

"Greek and medieval references to "evolution" use it as a descriptive term for a state of nature, in which everything in nature has a certain order or purpose."[1]

People, who wanted to "take God out of the picture" and give credit to man instead of God, birthed and spread evolution: to diminish God, and to rid Him completely. **Today, we have Christians who are diminishing God's Word, sovereignty, and power by enabling evolutionary thoughts to be spoken in God's house.**

We have already spoken about the Gospel in chapter 6. We have seen many aspects of it. The main aspect I would like to focus on is sin. Some may ask, "Why would you want to talk about sin in a chapter that is supposed to be about the effects on the Gospel?" To the compromising Christian who believes in evolution, I would respond with a question, **"Why did Jesus *have* to die?"**

Why Jesus *Had* to Die

I am going to make a bold statement here:

> **"Our Salvation depends on Jesus being real, which means Adam (and Original Sin) must be real. Without a Literal Creation, a Literal Adam and Eve, a literal Original Sin, and a literal Physical Death through Sin, there is no need for a Savior! Jesus had to conquer death for the propitiation of our sin!"**

If we go back to the beginning, all the way back to the very first sin of Adam and Eve, we will see the extreme importance of that singular sin. How can a sin, especially that one, be so important in a conversation about the Gospel?

As we covered earlier in chapter 3, God warned Adam, in Genesis 2, that death would be the result of eating from the tree of knowledge of good and evil. We know that it referred to *both spiritual and physical* death, as we covered in chapter 5. Without that sin, there would be no death in the world. Without that sin, there would be no judgment of God that needed to be poured out. Without that sin, there would be no hell, and without that sin, there would be no need of the Savior Jesus Christ.

That's just the key. Sin. This entire argument revolves around that single moment in time, where Eve and Adam decided that they knew better than the Omnipotent God. That horrid moment that they bit into the fruit - how sweet it must have been! Imagine, the juices flowing down their chins and the enjoyment of that moment when they tasted something that was forbidden. Now imagine, the moments afterwards, the guilt settling in and the realization of what they had just done. Immediately, they realized they were naked and ashamed of it. Puritan Thomas Watson rightly preached, "What fools are they, who, for a drop of pleasure, drink a sea of wrath."

The Cursed Creation

Thanks to Adam and Eve, all of creation was thrown into corruption. Death did not happen until after that very first sin:

> "Therefore, just as sin came into the world through one man, and death through sin, and so death spread to all men because all sinned." (Romans 5:12)

> "For as in Adam all die, so also in Christ shall all be made alive." (1 Corinthians 15:22)

That first sin, which led to a certain death and corrupted all of creation, led to countless sins and countless sacrifices in the Old Testament. That sin nature continues with every new birth today, and death eventually occurs to every one of those births. In the end, God will rid everything of evil. This includes conquering death itself:

> "The last enemy to be destroyed is death." (1 Corinthians 15:26)

So if death is a foe to be defeated, how then does it make sense that God would use evolution (which operates on continual death) before sin? *It does not, because death is a result of sin!* We see that there is an enemy that God has defeated through the death, burial and resurrection of Jesus Christ. This enemy is death itself. **He will not share His glory with anyone, especially His enemies (death).** We also see that all of creation was set to destruction when we read:

> "For the creation was subjected to futility, not willingly, but because of him who subjected it, in hope." (Romans 8:20)

So we can clearly see that there was no death and no destruction before the original sin of Adam. If you are reading your

Bible accurately and taking Genesis *literally*, like you *literally* should, you would come to the clear and logical conclusion that this lie of evolution could not be true. Only God is true as stated here:

"Let God be true though every one were a liar," (Romans 3:4)

God is right, and the "pseudo" scientists of this day are not. Again I say, there could not have been any death before sin! If you, as a Christian, who does not start with sin as the starting point of death, then you are calling God a liar. The problem with evolution is it revolves around death. *Death, upon death, upon death.* The real crux of the issue is that in an evolutionary worldview, death occurred before original sin. That is exactly the opposite of God's Word. That is often the pattern of the teachings of the world- not only with evolution, but with many other things as well.

That was the entire premise to the birth of evolution - to remove God completely. It has been a systematic abolishing of God's Word. Today, we even have Christians trying to mix the Christian worldview with those of the world. But as we have seen, we cannot do that. We must choose one over the other. We cannot attempt to serve two masters, as we have been warned against in Matthew 6:24.

The next thing you have to fully comprehend is that Jesus is real and He is who He says He is. He is God the Son, and He did die and rise again on the third day according to scriptures. He is now seated at the right hand of the Father. Our salvation depends on Him being real, and our salvation depends on His sacrifice taking away our sins. Unlike the sacrifices of the Old Testament that just covered the sin, like the one that God did to make the clothes for Adam and Eve, Jesus completely atones for our sins if we repent and believe the Gospel (Mark 1:15).

False Teachers Everywhere

With all of this capitulation of the modern church, we must be aware of false teachers and heretical doctrines more than ever. They will come in and soothe us into the pit of hell with the deceptions from the "father of lies." We are strictly warned about this in several passages. Here are three I think are important to look at:

> "Yet because of false brothers secretly brought in-who slipped in to spy out our freedom that we have in Christ Jesus, so that they might bring us into slavery". (Galatians 2:4)

Here, we see that "brothers" were secretly brought in to spy on the Christians to try to bring them into the "slavery" of the world. This is exactly what happens when we try to mix God's truth with the worlds "truth."

> "See to it that no one takes you captive by philosophy and empty deceit, according to human tradition, according to the elemental spirits of the world, and not according to Christ." (Colossians 2:8)

We are clearly told not to fall for the world's philosophy and empty deceit - *lies*. It is human tradition and even "spirits" of this world. This is, as we see, the opposite of Christ.

> "For certain people have crept in unnoticed who long ago were designated for this condemnation, ungodly people, who pervert the grace of our God into sensuality and deny our only Master and Lord, Jesus Christ." (Jude 4)

Ungodly people try to pervert the grace we are given and the words of our God. The intent is to deny our Master, His Lordship, and His hand in creation.

Far too often, we are taken captive by the world's philosophy. It tries to make us look like fools since we do not agree with its ideas. It mocks God. However, one day there will be a reckoning. God will judge the world. The World will reap what it has sown.

> "Do not be deceived: God is not mocked, for whatever one sows, that will he also reap." (Galatians 6:7)

It must be understood that *any* attempt to mesh evolution with the Bible undermines the *entire* Bible. The Apostle John provides a sober reminder of the eternal consequences of doing such a thing.

> "I warn everyone who hears the words of the prophecy of this book: if anyone adds to them, God will add to him the plagues described in this book, and if anyone takes away from the words of the book of this prophecy, God will take away his share in the tree of life and in the holy city, which are described in this book." (Revelation 22:18-19)

Chapter Twelve

A Call to Repentance

What is Repentance?

In this chapter, we will look at the issue of Biblical repentance in more depth. As you may recall from chapter 6, repentance is a change of mind and heart regarding sin. We once loved our sin. We used to dive into it. Now, even though we still sin, we hate our sin. That is the type of repentance a Christian has. God grants you the ability to repent- we must understand that repentance is a gift from Him, according to Scripture:

> " And the Lord's servant must not be quarrelsome but kind to everyone, able to teach, patiently enduring evil, correcting his opponents with gentleness. God may perhaps grant them repentance leading to a knowledge of the truth." (2 Timothy 2:24-25)

We see here clearly that God grants repentance leading to knowledge of the truth.

> "God exalted him at his right hand as Leader and Savior, to give repentance to Israel and forgiveness of sins." (Acts 5:31)

Repentance again is given.

> "Or do you presume on the riches of his kindness and forbearance and patience, not knowing that God's kindness is meant to lead you to repentance?" (Romans 2:4)

God's kindness is supposed to lead you to repentance - which *He alone* can give.

There are also many other verses that point to the fact that repentance is from God. The bottom line is that we can never make a claim to our own salvation. If we have ever done this, we need to fall on our faces and call out for forgiveness. We know that works do not save us:

> "For by grace you have been saved through faith. And this is not your own doing; it is the gift of God, not a result of works, so that no one may boast." (Ephesians 2:8-9)

So if any one makes a claim to any part of his own salvation, he must be called out, in love, to repent. **God gets the glory for every part of our salvation**. We have done nothing but sin against Him in our total depravity.

Is Repentance Considered a "Work?"

No! Many false teachers have preached incorrectly in regards to this. They have said that repentance is either a work or it is unnecessary. Yet, from Genesis to Revelation, we are told to repent of our sin. I have compiled a very short list to give an example. It is still hard to believe that someone would say that it is a work because, just like faith, it is a gift from God (Acts 11:18, 2 Timothy 2:25). It is also the result of the conviction of sin and faith in Christ. It doesn't matter which one comes first because they both must happen for salvation to take place. The following verses are examples of repentance in Scripture:

> "The time is fulfilled, and the kingdom of God is at hand; repent and believe in the gospel." (Mark 1:15)

"And Jesus answered them, "Those who are well have no need of a physician, but those who are sick. I have not come to call the righteous, but sinners to repentance." (Luke 5:31-32)

A partial listing: Matthew 3:8, Matthew 4:17, Matthew 21:32 (NIV), Mark 1:15, Luke 5:31-32, Luke 17:3-4, Luke 24:46-48, Acts 3:18-19, Acts 5:31, Acts 11:18, Acts 20:21, Romans 2:4, 2 Corinthians 7:9-10, 2 Timothy 2:25, Revelation 2:5, and Revelation 3:3.

False Gospels that Require Repentance

Many professing Christians in America today have no idea what the true Gospel is. Even worse, they have settled for a different gospel. Here are just two of the very strict Biblical warnings and prophecies about what will be happening in the church due to that:

"But even if we or an angel from heaven should preach to you a gospel contrary to the one we preached to you, let him be accursed." (Galatians 1:8)

"For the time is coming when people will not endure sound teaching, but having itching ears they will accumulate for themselves teachers to suit their own passions, and will turn away from listening to the truth and wander off into myths." (2 Timothy 4: 3-4)

We see that we are strictly told not to believe in another gospel. Not only do our "everyday sins" require repentance, but also does a belief in false Gospels.

The "Love" Gospel

I am sure you have heard, "God is a God of love." These false teachers make the Creator of the universe, the One who has laid the foundation of the world, into a sin-ignoring, all-about-the-love, "hippie savior" who contradicts himself in many ways. This is a fundamental re-definition of "love." This false gospel defines love as God spoiling human beings by ignoring their sin and giving people whatever they want. This makes God submissive to humans when it should be humans submitting to God. How dare we mar the name of Jesus Christ! We cannot describe a single person adequately by just naming one attribute of said person- we must name many attributes to just start to paint the picture correctly. God is indeed love, but let us not forget that He is also wrathful, just, omnipotent, and omniscient, among other traits. Many will go straight to John 3:16a and say, "See, it's all about love," but they forget the rest of the verse,

> "For God so loved the world, that he gave his only Son, that whoever believes in him should not perish but have eternal life." (John 3:16)

Did you see what it says? He gave His son... Why? Sin requires a payment. Jesus was God in the flesh and it took an eternal fine-payer to be able to pay the eternal fine.

Throughout the NT, God's love for us is always connected to His Son's sacrifice on the cross for our sins. The love is not in regards to us being loved in our sin; it is regarding the propitiation for them. If we are not born again, then the wrath of God remains on us. **He does not excuse sin because of love.**

> "Whoever believes in the Son has eternal life; whoever does not obey the Son shall not see life, but the wrath of God remains on him." (John 3:36)

He punishes sin because of love. Hell is just punishment because we sin against an eternal God who requires an eternal punishment. God loves us very much, and **He will and can save us, but it is on His terms and not ours.**

The Prosperity Gospel

We have already discussed why we should not believe another Gospel, but there is another facet of this. There has been more false teaching that has risen up. According to this prosperity gospel, God wants you to have your best life now. But, if you do not have it, it is because you do not have enough faith or because you need to give God permission to bless you.

There are far too many issues with this for me to cover here, but let us just say this: throughout the New Testament, God does not promise us riches and other blessings. Instead, He promises persecution and hate from unbelievers.

There are many solid Christians around the world who live in poverty. This aberrant teaching is unbiblical and will lead many down the wrong path. If you are in a church that teaches either of these false doctrines, do not walk away- run! Cling to God's Word and the truth in it.

> "Enter by the narrow gate. For the gate is wide and the way is easy that leads to destruction, and those who enter by it are many. For the gate is narrow and the way is hard that leads to life, and those who find it are few." (Matthew 7:13-14)

The "Works" Gospel

We have also addressed the true Gospel in chapter 6, and that *we are saved by grace, through faith, in Christ alone* (Ephesians 2:8-9). No amount of good works can erase our bad deeds and the penalty we owe. *This is the one thing that makes Christianity different from **all** other religions.* Every (false) religion on this planet has a "works" component, including: Roman Catholicism, Mormonism, Judaism, and Islam. Only Biblical Christianity is based not on anything that we do; instead, it is based on what Christ already did. As John Piper has said, "The gospel is not a help-wanted ad. It is a help-available ad."

The False Gospel involving Evolution

As we must look to Scripture for guidance on every issue, we must settle on some things that will embolden us in our stance on creation.

1. God is right!

"By no means! Let God be true though every one were a liar, as it is written, 'That you may be justified in your words, and prevail when you are judged.'"(Romans 3:4)

2. God will not be mocked!"

"Do not be deceived: God is not mocked, for whatever one sows, that will he also reap." (Galatians 6:7)

Having a true understanding of what sin is can really help us when we think about the issue of this false gospel of evolution and creation. If we have settled on the fact that **God has spoken and He cannot lie**, then it is a sin to say anything other than what He has stated in His word. Many try to turn this conversation of creation and evolution into a miry pit of straw man arguments and try to cause you not to be able to see through the fog of their pseudo-intellectualism (Romans 1:22). Remember that the clarity of God's word *will and should* guide us through these conversations on this subject (Psalm 119:105).

I really like the way Dr. Martyn Lloyd-Jones says it here:

"... we shall never have an adequate conception of the greatness of this salvation unless we realize something at any rate of what we were before this mighty power took hold of us, unless we realize what we would still be if God had not intervened in our lives and had rescued us. In other words, we must realize the depth of sin, what sin really means, and what it has done to the human race."[1]

Plainly put, professing Christians straying away from what the Bible says about any subject is apostasy. Once a Christian has heard the truth - which includes Genesis and the creation account - he no longer has the excuse of ignorance.

The main issue truly lies in how you view God's word. If you stand on *Sola Scriptura*, then you have to submit and agree with God's word in its entirety. When you say that you do agree with it and that it is inerrant and infallible, you cannot then say anything that opposes it. You certainly cannot use some of the science out there that contradicts Scripture. Otherwise, you are **double-minded** according to the Bible.

The Call

Repentance is a change of mind and direction. It is recognition that what you have done or what you are doing is wrong and sinful. It involves apologizing to God, crying out for forgiveness and turning away from your sin and turning to God. It is a complete abandoning of your love of sin.

Crying out to God in repentance is necessary not only for sin, but also if you have believed in any false doctrine. If you: have believed that evolution can be mixed in with Christianity, have not studied to show yourself approved (2 Timothy 2:15), have been ashamed of what the Bible says, or have wanted to be friends with the world, **then you must repent**. Cry out to God for forgiveness and He will forgive you. In 1 John, it says:

"If we confess our sins, he is faithful and just to forgive us our sins and to cleanse us from all unrighteousness." (1 John 1:19)

Take comfort, Christian. God loves you, and He will forgive you. If you have not been faithful regarding God's Word on creation, I urge you to repent.

Why Do Most Pastors Not Stand Firm on Creation?

There is a lack of action from many pulpits. There could be many reasons, and I do not presume to know anyone's heart. I do, however, know some excuses I have heard and some possible reasons. As a pastor, you are responsible for leading the flock Christ gave you. We know leaders are under more scrutiny than the flock (James 3:1). Pastors must take their position seriously and with soberness; yet, many end up not taking firm stances on tough issues. A large number of pastors are worried about tithe money and sheer numbers of attendees. They end up running the church as a business and worry about what they say from the pulpit to keep the lights on and seats filled instead of preaching God's Word in its entirety, including creation. Others refuse to take a stance because they do not believe that it is important. Even more, others want to be friends with everyone, something that is explicitly spoken against in James:

> "You adulterous people! Do you not know that friendship with the world is enmity with God? Therefore whoever wishes to be a friend of the world makes himself an enemy of God." (James 4:4)

Why Do Most Christians Not Stand Firm on Creation?

Many of the reasons pastors refuse to take a firm stance on creation are the same for the flock. A few have said it is not that big of a deal because it is not a salvific issue. Well, in a sense, they are right because you can be saved and believe in the lie of evolution. If, however, you have been made aware of the seriousness of the issue, and the truth of God's word, then you are going against what God has said and you are guilty of saying that God does not tell the truth. **If you give an inch in Genesis, you will more than likely give a mile everywhere else.** When you ask the question, "Why can the world's explanation not be mixed into the Bible," then you need to start asking where you stand with God. Others, who do not study to search these things out, may not know we are told to study to show our selves approved,

"Do your best to present yourself to God as one approved, a worker who has no need to be ashamed, rightly handling the word of truth."(2 Timothy 2:15)

They may also be afraid or ashamed to say what they truly believe because they think they will be looked down upon from the world. Let us just look at what Jesus said about us being ashamed of His words:

"For whoever is ashamed of me and of my words, of him will the Son of Man be ashamed when he comes in his glory and the glory of the Father and of the holy angels." (Luke 9:26)

As we see from this verse, we should never be ashamed of what God has said.

No Action From the Body of Christ as a Whole

When we think of the body, we must think of the many different denominations. If we agree on the main points, then we can count them as brothers and sisters in Christ. The lack of action from the body, I believe, is due to the lack of unity between the denominations. We have all been so consumed with proving each other wrong instead of proclaiming the word of God boldly and unashamedly. Remember Jesus wants us to be unified as a body. We are all different, but we are all His. Remember that Jesus prayed for us to be unified:

"That they may all be one, just as you, Father, are in me, and I in you, that they also may be in us, so that the world may believe that you have sent me." (John 17:21)

Jesus says unity is a powerful testimony to the world that God sent Him. Jesus even states unity is used by God to bring salvation to people. The world sits by and laughs at us and scoffs at God because His children have such a hard time getting along. There is no need for such disunity of the church. It is sinful.

Chapter Thirteen

BASIC CHALLENGES – AGE OF THE EARTH

The Age of the Earth

Most Christians acknowledge the authenticity of the genealogies in the Bible. From Adam to Abraham to Jacob, the Bible gives us an unbroken chronogenealogy. This period of time is about 2000 years. From Jacob to Jesus, historical details in the Bible give us a period of about 2000 years as well, with a possible error of only about 1%. Those 4000 years from Adam to Jesus, plus the historical record of 2000 years from Jesus to present time, gives us an age of about 6000 years from Adam to now.[1]

Because the Bible's chronogenealogies, Biblical historical details, and secular historical records **solidly** provide the proof that there was a total of about 6000 years from Adam to now, compromising or ill-informed Christians (including many pastors) have come up with two main methods to try and fit millions/billions years into the Bible.

- The Day-age theory
- The Gap Theory

Both of these methods attempt to insert the millions/billions of years into the 6 days of creation, since they cannot be put elsewhere!

The Day-Age Theory[2]

The Hebrew word, "yom," usually translates into the English word, "day." Depending on the context of the sentence, it can also be translated as "era." The word "yom" is used many times in the Bible; in fact, it is used 2301 times in the OT! In every one of those instances, besides the beginning of Genesis, the translation of "yom" is understood to be either "day" or "era" because of the context of the sentence.

Because it can be solidly shown that about 6000 years have elapsed since the birth of Adam on the 6th day, the day-age theory attempts to translate the "yom" into an "era" of time. Thus, each of the 6 "eras" would translate to millions/billions of years. The interesting thing is that the only place that the translation of "yom" is questioned is in the beginning of Genesis. Why? What are the rules that are used to understand the context surrounding the word "yom?" Then, what would happen if those same rules were used to understand the meaning of the word "yom" in the beginning of Genesis?

We know from the Bible and other Hebrew writings that the word "yom" always means a 24-hour day when it appears with:

- The word: morning,
- The word: evening, or
- A number

These "rules" are universally used to translate "yom" to "day" throughout the OT, but only *after* the beginning of Genesis. What, then, do we read in Genesis 1?

- "God called the light 'day' and the darkness he called 'night.' And there was evening, and there was morning – the first day." (Genesis 1:5)
- "And there was evening, and there was morning – the second day." (Genesis 1:8b)
- "And there was evening, and there was morning – the third day." (Genesis 1:13)
- "And there was evening, and there was morning – the fourth day." (Genesis 1:19)
- "And there was evening, and there was morning – the fifth day." (Genesis 1: 23)
- "And there was evening, and there was morning – the sixth day." (Genesis 1:31b)

When we apply these rules to Genesis 1, the word "yom" certainly translates to "day" for each of the 6 days! It does not translate to "era"! The day-age theorists cannot be right!

If that is not enough to convince a compromising Christian, then ask him what Exodus 20 is discussing. If he does not know what it is, kindly tell him that it is the passage in which God gave the 10 commandments to Moses. Ask Him if he believes in the 10 commandments. Then read him the following passage:

"Remember the Sabbath day, to keep it holy. Six days you shall labor, and do all your work, but the seventh day is a Sabbath to the Lord your God. On it you shall not do any work, you, or your son, or your daughter, your male servant, or your female servant, or your livestock, or the sojourner who is within your gates. For in six days the Lord made heaven and earth, the sea, and all that is in them, and rested on the seventh day. Therefore the Lord blessed the Sabbath day and made it holy." (Exodus 20: 8-11)

God not only spoke these to Moses, but He also inscribed them Himself onto the stone tablets. In the fourth commandment, God said that we would work for six days and rest on the seventh. This would make no sense if each day were actually a long era of time. **It is clear that God certainly spoke of a week being comprised of seven 24-hour days**.

Gap Theory (Ruin-Reconstruction)[2]

The gap theory, sometimes called the ruin-reconstruction theory, is another popular way of trying to force millions-to-billions of years into the Bible. For the same reason as the day-age theory, this theory was concocted because it can be solidly shown that about 6000 years have elapsed since the birth of Adam on the 6th day. Unlike the day-age theory where the length of the days of creation are questioned, the gap theory instead involves a supposed large gap of time between Genesis 1:1 and Genesis 1:2.

> **"In the beginning, God created the heavens and the earth."**
> **(Genesis 1:1)**
>
> (supposed millions/billions of years occur here)
>
> **"The earth was without form and void, and darkness was over the face of the deep"**
> **(Genesis 1:2)**

There are many different ideas as to what supposedly happened in this "gap" of time; it usually involves an original creation that God judged and destroyed. Thus, Genesis 1:2 would be the start of God re-creating everything on the earth. Most versions of the gap theory place millions of years of geologic time (including billions of fossil animals) in between these first two verses of Genesis. That would mean death, pain, killing, disease, thorns, struggle, suffering, and extinction would have occurred in this "gap." However, this presents a major problem – there was **no death before sin**. While humans would certainly not have died before Adam's sin about 6000 years ago, we similarly could not have a fossil record full of animal death (fossils). Yet, the fossil record that evolutionists believe to be millions/billions of years old is full of dead organisms, thorns and thistles, and fossilized bones with cancer! Also, keep in mind that while the gap theory teaches of a re-creation, this is not mentioned anywhere in the Bible.

Catastrophic Problems with Both Theories[2]

There are several catastrophic problems with the day-age and gap theories. While the compromising Christians have attempted to force millions-to-billions of years into the Bible, neither of these theories can stand up to the rest of Scripture. The three chief problems are:

> 1. Jesus's words in the Gospels puts Adam and Eve, the suffering of man, and Abel very near the beginning of creation. This makes no sense if there were millions/billions of years before Adam and Eve and original sin.

- "But from beginning of creation, 'God made them male and female.'" (Mark 10:6)
- "For in those days there will be such tribulation as has not been from the beginning of the creation that God created until now, and never will be." (Mark 13:19)
- "So that the blood of all the prophets, shed from the foundation of the world, may be charged against this generation, from the blood of Abel to the blood of Zechariah..." (Luke 11:50-51)

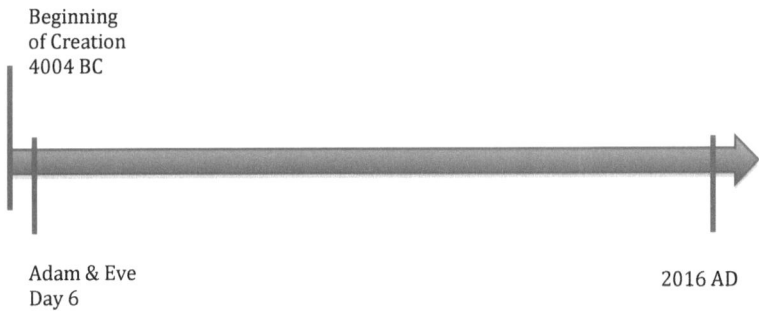

Literal Interpretation: 6,000 Year-old Earth

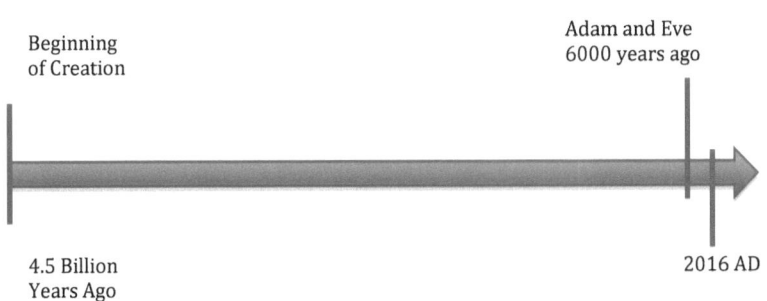

Non-Literal Interpretation: 4.5 Billion Year-old Earth

If almost 4.5 billion years have passed before Adam and Eve lived on the Earth and sinned, then they would not be very near the beginning of Creation. This makes no sense regarding the Word and the authority of Scripture.

2. The Bible clearly teaches that physical death is a result of original sin, committed by Adam and Eve. For either of these theories to be correct, it would have to presuppose death before sin. This would clearly contradict the Bible. ***The defense of "no death before sin" is the single greatest argument for a literal Genesis. This destroys every alternate theory of origins.***

3. The day-age and gap theories would make God's "very good" in Genesis 1:31 include death and destruction. This obviously contradicts Scripture as well.

It is clear that neither of these theories are correct- Genesis is certainly meant to be taken literally. God created six in 24-hour literal days, about 6000 years ago. If that sounds crazy, that only goes to show the effectiveness of brainwashing in our government schools in the last 100 years.

Chapter Fourteen

Basic Challenges- Part One

Is Genesis written to be poetic, rather than a historical narrative?[1]

This challenge typically comes from the "old earth" crowd who claim to believe in Biblical authority, but they deny the Biblical creation account as God has it recorded in Scripture. Rather than accepting the fact that Genesis easily reads as historical narrative, they attempt to add millions to billions of years into the first six days of creation by fabricating the idea that Genesis is actually written to be poetic. This means that the "old earth" Christians can interpret Genesis in the way that they wish to! Yet, we have great reason to believe that Genesis was written as historical narrative – everything was created in six, twenty-four hour, literal days.

The New Testament treats Genesis 1–11 as historical narrative. At least 25 New Testament passages refer directly to the early chapters of Genesis, and they are always treated as real history. Jesus cited Genesis 1 and Genesis 2 in response to a question about divorce (Matthew 19:4-6; Mark 10:6-9). Paul referenced Genesis 2-3 in Romans 5:12-19; 1 Corinthians 15:20-22, 45-47; 2 Corinthians 11:3; and 1 Timothy 2:13-14. The death of Abel recorded in Genesis 4 is mentioned by Jesus in Luke 11:51. The Flood (Genesis 6-9) is confirmed as historical by Jesus (Matthew 24:37-39) and Peter (2 Peter 2:4-9, 3:6), and in Luke 17:26-29, Jesus mentioned the Flood in the same context as he did the account of Lot and Sodom (Genesis 19). Finally, in Luke's genealogy of Christ, he includes 20 names found in the genealogies of Genesis 5 and 11 (Luke 3:34-38).

Several word studies have been done on the early chapters of Genesis, including ones completed by ICR and CMI. Every study shows that the words used point to a narrative and not poetry. One particular study, *Statistical Determination of Genre in Biblical Hebrew: Evidence for an Historical Reading of Genesis 1:1-2:3*, was completed by Steven W. Boyd, Ph.D. As stated in the study, " The logistic regression model calculates the probability that a text is a narrative. For Genesis 1:1–2:3, this probability is between 0.999942 and 0.999987 at a 99.5% confidence level. Thus, we conclude with statistical certainty that this text is narrative, not poetry."

If death only occurred after original sin, then what about plants? God said that he gave Adam and Eve and the animals all of the plants to eat.[2]

This question is meant to challenge the meaning of the word death in Genesis 2:17. Biblical creationists correctly understand that both spiritual and physical death only occurred as a result of original sin. Thus, humans (and animals) did not eat meat before the Fall in Genesis 3, as an animal would have to be killed to be able to eat meat. People ask this question to make the point that plants, which served as food, must have "died" before the fall. Do plants actually die?

The Bible makes a clear distinction between the status of plants and animals. People and animals are described in Genesis as having, or being, *nephesh* (Hebrew)—see Genesis 1:20-21, 24 where *nephesh chayyah* is translated "living creatures," and Genesis 2:7 where Adam became a "living soul" (*nephesh chayyah*).

Nephesh chayyah (living being/soul) is used in the Bible to describe sea creatures, land animals, birds, and man. *Nephesh* is never used to refer to plants. Therefore, the fact that plants were eaten before the Fall does not violate what God said in Genesis 2:17 regarding death!

What about the Fossil Record?[2]

Our fossil record consists of billions of dead things, exquisitely preserved, all over the world. Furthermore, it is full of complex, distinct organisms.

1. How do fossils form?

    - Is it from dead things collecting on the ground, and then sediment slowly covering up those dead things over many years? Many of those rock layers that the fossils are contained in supposedly took millions of years to form.
    - Or, do fossils generally form by a QUICK and DEEP burial of organisms that were buried either alive or very recently dead? Think about an animal that dies in the field. What happens to it? Once scavengers, bacteria, and the decay process gets ahold of the carcass, the body and bones disappear in a matter of weeks to months!

Obviously, it is the latter explanation! And this is explained very easily by a rapid burial from a worldwide flood!

2. Where are all of the "transitions?"

    - In his book, *On the Origin of Species by Means of Natural Selection, or the Preservation of Favoured Races in the Struggle for Life*, Charles Darwin[2] admitted, "Why then is not every geological formation and every stratum full of such intermediate links? Geology assuredly does not reveal any such finely graduated organic chain; and this, perhaps, is the most obvious and gravest objection which can be urged against my theory."
    - Where are the millions of transitions between the Cambrian and Precambrian periods? There are no life forms that go from single cell, to multiple cells, to hundreds of cells, etc., ... all the way up to complex life forms. There is only the *sudden* appearance of complex creatures created after their own kind. The evolutionists try to explain this away as the "Cambrian Explosion." They say that for some reason, not much evolving was happening for many years, then all of a

sudden- "Bam!" New complex creatures started to pop-up all over the earth for no apparent reason.
- Thus, scientists only claim to have a few transitional life forms (which creationists dispute), when *there should be millions of them!*
- As one example, scientists claim that modern day birds evolved from dinosaurs. We see many dinosaurs in the fossil record. The same goes for birds. How many different missing links are needed to have dinosaurs evolve into birds? Our fossil record should contain many fossils of each of the myriad of supposed transitions. Yet, there are no good examples (scientists have a few that they claim are transitions, but that still does not answer the question at hand).

The real explanation – God created everything to reproduce after its own kind. There are no transitional fossils! The supposed "Cambrian Explosion" is just the proof of a worldwide flood that buried billions of complex organisms quickly!

Who was Cain's Wife?[3]

Whether they believe it to be true or not, most people understand the Biblical teaching that all people came from Adam and Eve. However, for this to be the case, who did Cain marry? The only human females that would be around would be his family members! Because of current laws and societal understanding that we are not to marry our close relatives, people who ask this question are ready to ridicule you in their "gotcha" moment. You either must say that there were other non-related humans around, thereby undermining the Bible, or you must admit to who Cain married - his sister or very close relative! We know from the genealogies in Genesis 5:

> "The days of Adam after he fathered Seth were 800 years; and he had other sons and daughters." (Genesis 5:4)

Thus, he would have married his sister (most likely) or a close relative, like a niece. Why was this not a problem?

1. All humans are related.
2. Abraham was married to his half sister. It was not until much later that God instructed the Israelites not to marry their close relatives – a principle we still follow today.
3. When close relatives marry today, there is an increased likelihood of deformities in the offspring due to mutations that have accumulated in the human race since Adam's sin. The amount of genetic mutations, as well as the diversity of mutations, has been increasing in every generation since Adam and Eve. However, the closer the relatives, the more likelihood that such people will have the same genetic mutations (mistakes). If close relatives were to marry today, there would be a higher likelihood of their children to receive a "double hit" on the genetic mutation (the same mutation from both mom and dad) to cause a disease or deformity.
4. However, the farther back in history one goes (towards the Fall of Adam), the less mutations there were. At the time of Adam and Eve's children, there would have been very little mutations in the human genome – thus close relatives could marry and not have to worry about their children receiving the "double hit" on the mutation.

How did Adam name all the animals?[3]

When reading Genesis 1 & Genesis 2, it is apparent that Adam named the animals sometime on Day 6. Scoffers point out that if Genesis is literal history and the days of creation are each twenty-four hour literal days, then it would have been impossible for Adam to accomplish this task. Because of the large amount of animals that we see today, their conclusion is that Genesis is not to be taken literally.

Yet, in reading Genesis 2 properly and using some simple mathematics, we can answer this question:

> "Now the Lord God had formed out of the ground all the wild animals and all the birds in the sky. He brought them to the man to see what he would name them; and whatever the man called each living creature, that was its name. So the man gave names to all the livestock, the birds in the sky and all the wild animals." (Genesis 2:19-20)

Notice that amphibians, sea animals, and arthropods are not mentioned in this passage, which make up over 98% of all living organisms. It is estimated that the number of creatures that Adam actually had to name was in the hundreds, not thousands or more. The best estimate of kinds (corresponding to modern "families") comes out to *less than 200 animals* that Adam had to name. If he named 200 animals at a rate of 1 animal every 15 seconds, Adam could have named them all in less than an hour.

How did Noah round up all of the animals on the Ark?[3]

The answer is simple - He did not have to! God sent the animals to him!

> "And of every living thing of all flesh you shall bring two of every sort into the ark, to keep them alive with you; they shall be male and female. Of the birds after their kind, of animals after their kind, and of every creeping thing of the earth after its kind, two of every kind will come to you to keep them alive" (Genesis 6:19-20)

How did all of the animals fit on the ark?[3]

Genesis describes the Ark in 3 verses:

6:14 – "Make yourself an ark (tebah) of gopher wood; make rooms in the ark, and cover it inside and outside with pitch
6:15 – "And this is how you shall make it: The length of the ark shall be three hundred cubits, its width fifty cubits, and its height thirty cubits." (450ft x 75ft x 45 ft)

6:16 – "You shall make a window for the ark, and you shall finish it to a cubit from above; and set the door of the ark in its side. You shall make it with lower, second, and third decks."

John Woodmorappe, who wrote the book *Noah's Ark: A Feasibility Study*[3], suggests that, at most, 16,000 animals were all that were needed to be on the Ark to preserve all created **kinds** (new studies show half of that).

- God only had the Ark carry air-breathing, land dwelling animals, creeping things, and winged animals (like birds). Aquatic life was not on board!
- The tremendous variety in species that we see today did not exist in the days of Noah. Only parent **kinds** were needed on board to repopulate the earth.
- Woodamorappe concluded that less than half of the area on the Ark was needed for the animals and their enclosures.

Some more recent studies regarding original *kinds* of animals on the Ark suggest that only about 30% of the Ark's space was taken up by the animals. The rest of Noah's Ark would have been open for the living quarters for Noah's family (8 people: Noah, his wife, their 3 kids, and their 3 kids' wives) and needed food and supplies.

Are dinosaurs real?

Yes! All land-dwelling creatures were made on Day 6, same day as Adam and Eve - this included the dinosaurs! Because the word "dinosaur" wasn't invented until 1841, more than ten years after the words "computer" and "rocketship," it doesn't appear in the Bible. The word used in the Bible for dinosaur is "dragon," which appears many times in Scripture, especially in the book of Job.

Were dinosaurs on the Ark?[3]

Because of movies and cartoons, we have a belief that all dinosaurs are huge. How could they possibly fit on the Ark? We know from Scripture that two of every kind of land animal boarded the Ark (and seven of some kinds). We know that dinosaurs were created on Day 6 and must have been included on the Ark, but did they go extinct before the Flood? The answer is no, for a few reasons:

1. The description of "Behemoth" in chapter 40 of the book of Job only fits something like a sauropod dinosaur (Job lived after the flood). Thus, dinosaurs survived the flood by being on the Ark.
2. The flood layers laid all over the earth (fossil record) contain dinosaur fossils.
3. There are many post-flood accounts of living dinosaurs, including some of the writings of Marco Polo and Pliny the Elder.

Dinosaurs could easily fit on the Ark. We know that:

1. Most dinosaurs were not large; most scientists agree that the average size of a dinosaur is the size of a sheep.
2. God most likely brought Noah two young adult dinosaurs of each kind.

3. We know that dinosaurs hatched from eggs, and since the largest egg is only the size of a rugby ball, even the largest dinosaur would have been small enough to fit in the egg.
4. There was probably only about 50 different kinds of dinosaurs, meaning that only 100 or so dinosaurs were needed on the Ark.

Aren't we almost the same as a chimp genetically? I thought our genomes are over 99% the same...[5]

No, we are not almost the same as a chimpanzee. "Evolutionary biased" scientists arrive at the 99% similarity between the genomes of the chimp and human. By assuming that chimpanzees and humans are evolutionarily related, theses scientists "match only the most similar regions (of the genomes), and ignore the rest (of the genomes)." Within these "alignment regions," the similarity is as high as 99%. However, when the entirety of the two genomes is compared, there is only about a 72% similarity between them. "Conservatively, if the human genome is over 3,000,000,000 (3 billion) base pairs," and the difference is 28%, there would be 840,000,000 (840 million) base pair differences between chimps and humans! For the scientists that hold to evolution, this difference in genomes cannot be accounted for in billions of years of time, let alone the few million years that supposedly separate chimps from humans.[5]

The best way to explain these hot zones is with the following example. Suppose that you want to compare two books to see how similar they are to one another. While "Book A" is a 100 page story all about recycling, "Book B" is a 120 page story about the insect world. Both books are written in English and are on a second-grade level of understanding, so they both use similar easy-to-understand words throughout. On page 45, "Book A" has the words, "it is a green bag." On page 56 of "Book B," you can find the words, "it is a green bug." Now, instead of taking the books in their entirety to compare their overall similarity, the evaluator only uses those two statements. The evaluator then sees that 16 of 17 characters (letters and spaces) are the same. Within this "alignment zone," the books are 16/17, or 94.1% the same. He would then (falsely) proclaim that the books are 94.1% similar overall.

Using the above example, the bias of the evolutionary scientists can easily be seen when attempting to evaluate the similarities between the chimp and human genomes. This bias has led to the widespread misperception of the 99% similarity between humans and chimps.

Who made God?

This is often an objection given by the unbelieving scoffer. They think that they won the debate when this question is asked. We understand that the universe has only 3 possible origins:

1. It came from nothing.
(This is impossible and goes against a known scientific law, which states that matter can neither be created nor destroyed.)

2. It created itself.
(This is impossible because something that does not exist cannot make anything.)

3. Something created it.

We are left with #3 as the only possibility- something created the universe. Thus, unbelievers are left with the same issue as creationists- something must have always existed for the universe to be created. Philosophers call this the "uncaused cause." Everyone must appeal to an uncaused cause, since something cannot come from nothing. Therefore,

> Creationists believe, "In the beginning, God." Therefore, God created the universe. God had no cause because He has always existed.
>
> Evolutionists believe, "In the beginning, stuff." They must also believe that something has always existed.

It is apparent that evolutionists believe in religion, too. Theirs just does not make any sense!

Irreducible Complexity

Evolutionists always want to start to explain evolution by commencing with the "simple" single-celled organism. But how could all of the many different, non-living parts of a cell evolve at the exact same time and come together into a living cell? They could not. Therefore, every one of the non-living parts must have magically formed at the same time to be the perfect size, shape, material, among having characteristics, to be able to come together into a living cell. This is impossible.

Think about the most commonly used example: the mousetrap. The mousetrap is made up of 5 integral parts: a catch, a spring, a hammer, a holding bar and a foundation. A mousetrap would not work without all 5 perfectly designed pieces put together perfectly. Now, imagine a living cell that has many integral pieces that must be perfectly designed and put together perfectly. Because each of the parts is non-living, they could not have evolved into the perfectly designed shapes that they are. Thus, a cell is irreducibly complex because each part of the must have magically been made exactly how it is and put together perfectly for the living cell to be formed. **This is not statistically possible!**

This problem of irreducible complexity can be applied to many other parts of the body. One example is this: What evolved first -the blood or the heart that pumps it, which must rely on the blood to even work? The circulatory system has parts that make it irreducibly complex.

There are many examples of irreducible complexity in the human body, as well as throughout God's entire creation! Even the simplest cells require DNA that is irreducibly complex. It just shows that all of creation is designed; we humans are fearfully and wonderfully made! (Psalm 139:14)

The Miller-Urey Experiment

Some evolutionists, when trying to explain how life could form from non-life, often cite the Miller-Urey experiment. This experiment attempted to prove that belief, even though it has a number of catastrophic problems. Yet, without citing this experiment, they have no way to explain how a single-celled organism could form, which then magically evolved into everything that we see today.

We as creationists often let evolutionists get away with this. They always want to start to explain evolution starting with the "simple" single-celled organism. While we would argue that there is nothing simple about even one living cell that has millions of non-working parts, as explained in the "irreducible complexity" segment above, this experiment deserves at least a mention.

"The Miller–Urey experiments involved filling a sealed glass apparatus with the gases that Oparin (A Russian biochemist who proposed that life evolved though gradual chemical evolution in a primordial soup) had speculated were necessary to form life—namely methane, ammonia and hydrogen (to mimic the conditions that they thought were in the early atmosphere) and water vapour (to simulate the ocean). Next, while a heating coil kept the water boiling, they struck the gases in the flask with a high-voltage (60,000 volts) tungsten spark-discharge device to simulate lightning. Below this was a water-cooled condenser that cooled and condensed the mixture, allowing it to fall into a water trap below... In time, trace amounts of several of the simplest biologically useful amino acids were formed—mostly glycine and alanine. The yield of glycine was a mere 1.05%, of alanine only 0.75% and the next most common amino acid produced amounted to only 0.026% of the total—so small as to be largely insignificant. In Miller's words, 'The total yield was small for the energy expended.' The side group for glycine is a lone hydrogen and for alanine, a simple methyl (–CH_3) group. After hundreds of replications and modifications using techniques similar to those employed in the original Miller–Urey experiments, scientists were able to produce only small amounts of less than half of the 20 amino acids required for life. The rest require much more complex synthesis conditions."[6]

There are many problems with this experiment:

1. Oxygen was left out of this experiment on purpose. If oxygen were present, it would have broken down the newly-formed amino acids as quickly as they would be created. Yet, for life to live, oxygen would need to be present from the beginning. In addition, the ozone, which is made up exclusively of oxygen) needed to be present from the beginning, otherwise life would have been baked to death! **This is an insurmountable problem: the amino acids cannot exist in the presence of oxygen, yet life would need oxygen to live and form the ozone to prevent the frying of life!**

2. Only a few of the 20 essential amino acids were made. **This cannot account for life!**

3. Equal quantities of left- and right-handed amino acids were produced in this experiment, yet **only left-handed amino acids are used in life! For life to have magically evolve, only the left-handed amino acids could be chosen!** What is interesting is that once a person dies, the left- handed amino acids revert back to a 50%-50% mix of left- and right-handed amino acids.

4. The irony is that, to make amino acids in this experiment, **it was designed** by humans! It didn't happen by chance!

The Cambrian Explosion

Evolutionists have a huge problem, as the intermediate species (transitional species) within the fossil record seem to be woefully deficient. The best that evolutionary scientists can do is point to a few possible examples, every one of which creationists would correctly dispute.

Precambrian rock contains very little fossil remains. The next rock layer, the Cambrian, "contains an impressive collection of diverse life-forms without identifiable ancestral forms. Dominated by marine invertebrates, Cambrian rock also contains the deepest vertebrate fossils. Therefore evolutionists ordinarily call it the 'Cambrian explosion' to describe the inexplicably abrupt evolutionary appearance of such great worldwide biodiversity."[7]

Why does it seem like there are no fossils of simple or intermediate species below the Cambrian layer, but then the Cambrian layer is full of complex organisms? Is it because the transitional fossils miraculously disappeared? No. It is because God created every complex kind of organism that we see today, and that was the result of them all being buried at about the same time during the worldwide flood in the time of Noah!

What about radiometric dating methods? (Excluding radiocarbon- C14)

Some scientists use radiometric-dating methods[8] to date rock layers in the millions of years. To understand this, imagine an hourglass. If we start an hourglass and return to it when there is exactly half of the sand on top and half of the sand on bottom, we would assume that 30 minutes of time have passed.

With radiometric dating methods, we have a parent atom that takes a certain amount of time to decay into the daughter atom. The half-life of that parent atom is the amount of time it takes for half of the parent atoms to decay into the daughter atom.

Thus, let us start with any amount of the parent atom, Uranium-238, and none of the daughter atom that it decays into, Lead-206. With a theoretical half-life of 4.5 billion years, if we would find a rock sample that had 50 atoms of U-238 in it, and 50 atoms of Lead-206 in it, we would assume that 4.5 billion years would have passed. If a rock sample had 25 atoms of U-238 and 75 atoms of Lead-206, we would assume 9 billion years, two half-lifes, would have passed. This is how radiometric dating of rocks theoretically works.

The problem with radiometric dating is that there are a lot of assumptions that are made, which can affect the results of the analysis significantly. This makes radiometric dating very unreliable. Some of these assumptions include:

1. Assuming the rock sample started with all parent and no daughter atoms present (even though no one can know what the initial conditions of the rock sample were).

2. Assuming that there was no contamination of the rock sample with either parent or daughter atoms.

3. Assuming that the rate of decay is always constant.

And many others...

These assumptions make the radiometric dating method very unreliable. This is shown in results of The Rate Project[9], which found that testing the same rock samples with different radiometric dating methods resulted in vastly different ages! A few examples include:

RATE Results: Bass Rapids diabase sill

Method	Age
Potassium-Argon	841.5 million years
Rubidium-Strontium	1060 million years
Uranium-Lead	1250 million years
Samarium-Neodymium	1379 million years

RATE Results: Cardenas Basalt

Method	Age
Potassium-Argon	516 million years
Rubidium-Strontium	1111 million years
Samarium-Neodymium	1588 million years

The two rock layers have radiometric dating methods that differ by 500 million and 1 billion years, respectively! There are problems with these methods!

One of the other great examples of problems with radiometric dating methods is concerning Mount St. Helens.

> "On May 18, 1980, a tremendous landslide on the northern side of Mount St. Helens in Washington State uncapped a violent volcanic eruption, completely altering the surrounding landscape. It is the most studied volcano in history and has reshaped thinking regarding catastrophic earth processes. The Institute for Creation Research has studied the volcano over the past three decades, conducting research that has provided a suite of informative lessons with broad-ranging implications... The mountain also provided a clear reason to distrust the reliability of radiometric dating. A new rock cap atop the mountain that formed after the 1980 eruption should have shown it to be on the order of tens of years. But standard analysis gave the totally incorrect date of 350,000 years. What other rocks have been dated incorrectly by following those standard dating protocols?"[10]

This is just one more obvious example of why we cannot trust the dates given by radiometric dating methods!

Chapter Fifteen

Basic Challenges- Part Two

Where did all of the different people groups come from?

This is a common question brought up by Biblical Creationist Christians and Old Earth Christians alike, as well as unbelievers. Because the Bible clearly teaches that all humans are descendants of Adam and Eve, all people must be of the same blood and same race – the human race! Yet, why do we see many different traits and varying shades of brown among people today? How did all of the different people groups that we see today arise from only Adam and Eve?

The background of what happened since the flood, and at the tower of Babel, is important to understand in answering this question. There were only 8 people who survived the flood: Noah, his wife, his three sons (Shem, Ham, and Japheth) and each of their wives. The repopulation of the earth after the flood originated solely from them.

Noah and his family were given clear instructions from God while coming off of the Ark:

> "And God blessed Noah and his sons and said to them, 'Be fruitful and multiply and fill the earth.'" (Genesis 9:1)

When we get to the historical account regarding the Tower of Babel, spoken of in Genesis 11, we see that the descendants of Noah didn't exactly follow the instruction above:

> "Now the whole earth had one language and the same words. And as people migrated from the east, they found a plain in the land of Shinar and settled there. And they said to one another, "Come, let us make bricks, and burn them

thoroughly." And they had brick for stone, and bitumen for mortar. Then they said, "Come, let us build ourselves a city and a tower with its top in the heavens, and let us make a name for ourselves, lest we be dispersed over the face of the whole earth." And the Lord came down to see the city and the tower, which the children of man had built. And the Lord said, "Behold, they are one people, and they have all one language, and this is only the beginning of what they will do. And nothing that they propose to do will now be impossible for them. Come, let us go down and there confuse their language, so that they may not understand one another's speech." So the Lord dispersed them from there over the face of all the earth, and they left off building the city. Therefore its name was called Babel, because there the Lord confused the language of all the earth. And from there the Lord dispersed them over the face of all the earth." (Genesis 11:1-9)

"Renowned chronologist Archbishop James Ussher placed the time of Babel at 106 years after the Flood, when Peleg was born."[1] We also know that Peleg was in the 4th generation after the Flood. So, only a short amount of time had passed from Noah and his family coming off of the Ark and receiving the above-mentioned command of Genesis 9:1, to the time of Peleg. In fact, according to the genealogies in Genesis 11, at least Noah and his son, Shem, eyewitnesses of the Flood, were living at the time of the dispersion at the tower of Babel!

Because of the scattering of people from the Tower of Babel, we can explain:

1. The formation of the different people groups,
2. The formation of different languages,
3. The Entrance of Human Religion,
4. The birth of the hundreds of different creation accounts and ark/flood legends around the globe. All human religions have followed what happened at the Tower of Babel in going their own way, inventing myths to replace God's account of creation and Noah's Flood.

The formation of people groups happened due to the fact that people were dispersed all over the face of the earth. Two factors play a role in this. First, as we learned earlier, there is a huge amount of genetic variation within the human kind. There is enough genetic variation among a husband and wife that they could have 10^{27} children before two of them would be genetically identical (excluding identical kids). That does not even consider the fact that there would have been much more variation between Adam and Eve!

Besides the normal variants within the genome, different "mutations that occurred in the different subpopulations after Babel," also explains how different physical traits would develop among the human race.[2] Those people groups, after being separated from the others, would cause different cultures to develop the slightly distinctive traits that we see today. And with the different people groups forming, new languages developed naturally.

By the moment of the dispersion, all of the different people groups would have had the same opportunity to possess the knowledge of Noah and his family. This means that they would have all left Babel with the same knowledge of the one true God, as well as the correct creation and flood accounts. This explains why there is at least some knowledge among cultures all over the earth today regarding creation and the flood. The fact that God preserved His Word through His chosen people, through the line of Seth, is how we know what the true accounts are. Ultimately, it is because of sin that caused the incorrect details to arise in the rest of the creation and flood accounts across the globe.

What about Distant Starlight?

This is the "distant starlight problem" in a nutshell. When a star is said to be a certain distance away from the earth, such as 13.5 billion light-years away, it means that the star's light, traveling at about 186,000 miles per second (the *average* speed of light), will not reach the earth for 13.5 billion years. If we can see a star here on earth, it theoretically should have taken 13.5 billion years to arrive.

Thus, the secular scientists conclude that the universe must be at least as old as the farthest star that we can see here on earth – 13.5 billion years.

There are some great resources (especially by Dr. Jason Lisle) that go over this topic in great detail. In a witnessing encounter, however, it is generally unnecessary to go into all of it. As creationists, we get accused of the fact that the universe cannot be only 6,000 years old when the starlight "proves" the universe is at least 13.5 billion years old. So, we are faced with the "problem" of proving how light got here much faster than "science" allows. The interesting thing is that the Big Bang model has the exact same problem- called "The Horizon Problem"- where light is assumed to have traveled faster than what we know to be the "speed of light."

> *In a nutshell, the Horizon problem is as follows*: "The temperature of the cosmic background radiation is essentially the same everywhere in the universe —in all directions (to a precision of 1 part in 100,000). However (according to big bang theorists), in the early universe, the temperature of the CMB (cosmic microwave background) radiation[7] *would have been very different at different places in space due to the random nature of the initial conditions*. These different regions could come to the same temperature if they were in close contact. More distant regions would come to equilibrium by exchanging radiation (i.e. light). The radiation would carry energy from warmer regions to cooler ones until they had the same temperature.
>
> *The problem is this*: even assuming the big bang timescale, there has not been enough time for light to travel between widely separated regions of space. So, how can the different regions of the current CMB have such precisely uniform temperatures if they have never communicated with each other? *This is a light-travel–time problem."* (Adapted from the article, *Light-travel time: a problem for the big bang,* by Dr. Jason Lisle, appearing on creation.com)[3]

The big bang theorists have the exact same "light travel problem" that the creationists do – they also acknowledge that light must have traveled faster than what is currently believed to be the speed of light!

Doesn't Radiocarbon (C-14) disprove the Bible?

No! It actually provides great support for a young earth! Many people falsely assert that radiocarbon dating proves that dinosaurs/ fossils/ the earth is millions to billions of years old. When that statement is made, the person is unknowingly admitting that he does not know how radiocarbon dating even works! However, a discussion regarding radiocarbon points directly to evidence for a young earth, just like the Bible teaches. Some facts about radiocarbon include:

1. It has a half-life = 5730 years
2. It is a reliable dating method up to *only 100,000 years.* (Some studies suggest its reliability as low as 60,000 years) After that, it cannot be detected.
3. If every atom in the earth were C-14, it would all have decayed away in less than 1 million years!
4. It is mostly used to date organic material, especially once-living things.
5. Radiocarbon cannot be used directly to date rocks. It can only be used to date organic material that is contained in the rock.
6. While it cannot date rocks directly, it can be used to put time constraints on some inorganic materials that could contain carbon, such as diamonds and coal.

Based on these facts, you cannot use radiocarbon to date things at millions of years! Anything that is "older" than 100,000 years should contain NO detectable radiocarbon. **Thus, if radiocarbon is detected in a fossil or another carbon-containing material, then it must be less than 100,000 years old!** Following are some carbon-containing substances and respective ages assigned by some scientists:

Coal Seams – At least tens of millions of years old
Diamonds – 1-3 billion years old
Dinosaur Bones – At least 65 million years old

What is interesting is that when C-14 dating is completed, every coal seam, diamond, and dinosaur fossil ever tested for radiocarbon has detectable amounts of C-14:

Coal Seams – Have detectable C14, must only be thousands of years old
Diamonds – Have detectable C14, must only be thousands of years old
Dinosaur Bones – Have detectable C14, must only be thousands of years old

It means that coal, diamonds, and dinosaurs *must* be less than 100,000 years old... they cannot be millions/billions of years old! Obviously, even an age of 100,000 years old is still higher than the 6000-year life of the earth. Articles by Answers in Genesis, Creation Ministries International, and Institute For Creation Research provide excellent explanations for this, especially regarding some of the assumptions that have to be made in the radiocarbon testing method. Radiocarbon remains a great test for an upper boundary of age (100,000 years old) and for relative dating among multiple substances, but it is not reliable for an exact age.

Vestigial organs

Over the years, if scientists did not see or realize a function of an organ, it was labeled as vestigial. Assuming that one believes in evolution, a common definition for a vestigial organ is an organ that was useful in an animal's evolutionary past, but is now useless or very close to useless. A few well-known examples are the:

1. Appendix: We have known for decades that it is a safe house that harbors useful bacteria for the gut. When someone gets diarrhea, and loses the beneficial bacteria of the gut, the appendix will replace the bacteria in the intestine. While a person can live without one (like so many of our other body parts, like eyes and hands), the appendix certainly serves a purpose!
2. Tailbone (coccyx): This is the focal point of the attachment of all the muscles that form the floor of the pelvic diaphragm. This diaphragm supports organs that reside in the pelvis (bladder, prostate gland or uterus, sigmoid colon. It certainly provides a lot of functions!

Some previously thought vestigial organs include: the tonsils, pineal gland, and the thymus. Like the examples above, these organs all proved to have a function as the human body was studied more over time.

There are some scientists, in trying to hold onto the concept of vestigial organs, who have tried to claim that some of these body parts, like the appendix or coccyx, should still be called vestigial because humans could live without them. This is where the "very close to useless" part of the definition above comes from. However, this does not make sense logically. By that reasoning, we should be able to call a number of our body parts vestigial, like the big toe. Even though it helps with our ability to balance and walk, we could technically live without it. Obviously, that idea is ludicrous. *We can be certain of one thing concerning so-called vestigial organs: it is quite obvious that there is not a single part of God's design for the body that is useless, even if we as humans do not know enough to understand their function!*

How could the first three days of creation be literal days if the sun was not created until day four?

Many people mistakenly believe that a day on earth is dependent on the sun. However, a 24-hour day on earth has nothing to do with the sun- it has everything to do with the earth's rotation on its own axis. The earth rotates on its axis 1 time in 24 hours. Because of the rotation of the earth and the tilt of its axis, we get daylight for a certain amount of time during the day, and then no sunlight (other than that reflected by the moon) for the balance of the 24-hour day.

Where did light come from if the sun wasn't created until day 4?

We know that from Genesis 1, light was created on the first day. The sun and moon were created on Day 4 to "govern" the light.

> "And God said, "Let there be light," and there was light…God called the light Day, and the darkness he called Night. And there was evening and there was morning, the first day." (Genesis 1:3-5)

> "And God said, 'Let there be lights in the expanse of the heavens to separate the day from the night. And let them be for signs and for seasons, and for days and years, and let them be lights in the expanse of the heavens to give light upon the earth.' And it was so. And God made the two great lights—the greater light to rule the day and the lesser light to rule the night—and the stars." (Genesis 1:14-16)

The plants, which were created on day 3, *did not need the sun as there was still light*. We cannot be dogmatic about the source of the light, but we do know from Scripture that *God is light*. Yet, we know that in the eternal state, the stars will be gone and God will be the light:

"And the city has no need of sun or moon to shine on it, for the glory of God gives it light, and its lamp is the Lamb." (Revelation 21:23)

"And night will be no more. They will need no light of lamp or sun, for the Lord God will be their light, and they will reign forever and ever." (Revelation 22:5)

Where did the dinosaurs come from?

Museums throughout the world display dinosaurs. What we are not seeing are the precursors to the dinosaurs. There should be millions of transitional species before they appeared on earth in the evolutionary timeline! Why aren't they in the museums? Are there even any precursors to the dinosaurs?

"The question of the origin of dinosaurs is one that has puzzled paleontologists for many years."
- The Illustrated Encyclopedia of Dinosaurs, Dr. David Norman, 1985

"Where did dinosaurs come from? That apparently simple question has been the subject of intense debate amongst scientists for over 150 years, …"
- The Natural History Museum Book of Dinosaurs, 1998

Just like all of the other complex organisms represented in the fossil record, it is obvious that the dinosaurs came into existence without any precursors. That is because God created them, like every other kind of organism, during the creation week!

Why is there death and suffering in the world?

As a common theme throughout the Bible, and spoken of throughout this book, *death is a result of sin!* God is not the author of death. According to the Bible, *death is the result of sin caused by*

Adam. God originally created a perfect creation. It included no death; in fact, He proclaimed everything to be "very good" in Genesis 1:31. All animals were created to be vegetarian; neither humans nor animals would kill animals for food, which was consistent with no death in His creation. As a result of the Fall, everything bad- death, disease, suffering, extinction, thorns and thistles- entered into the world. Animals started to kill and eat other animals after the Fall, and humans were given that ability after the Flood.

When we understand the correct Biblical framework of earth's history, the earth was created perfectly in six 24-hour literal days, about 6000 years ago. Adam and Eve, through original sin, brought death into the world. Death did not exist before that! **Tragedy in someone's life can only be explained properly with a correct Biblical worldview, where death is *only* a result of Adam and Eve's original sin!** All of the bad things in this life are the fault of humans, not God. God will eventually eradicate all tragedy, disease, and everything else bad after He returns. After all,

> "The last enemy to be destroyed is death." (1 Corinthians 15:26)

When a professing Christian compromises with the belief that the earth is millions/billions of years old, death must be incorrectly placed *before* original sin. God would have then used death in evolution to make every organism, including humans. However, if that is the case, then how can we explain tragedy? We would have to blame God for death. People would be able to be "mad at God" if he authored it. Obviously, this is not the case as sin and death would have had to commence before the Fall.

> "Let no one say when he is tempted, "I am being tempted by God," for God cannot be tempted with evil, and he himself tempts no one. But each person is tempted when he
> is lured and enticed by his own desire. Then desire when it has conceived gives birth to sin, and sin when it is fully grown brings forth death." (James 1:13-15)

Human sin brought death into the world – therefore, we as humans deserve the blame for all tragedy. But the curse is not just on humans - it is over the entire creation:

> "For the creation was subjected to futility, not willingly, but because of him who subjected it, in hope that the creation itself will be set free from its bondage to corruption and obtain the freedom of the glory of the children of God. For we know that *the whole creation* has been groaning together in the pains of childbirth until now. And not only the creation, but we ourselves, who have the firstfruits of the Spirit, groan inwardly as we wait eagerly for adoption as sons, the redemption of our bodies." (Romans 8:20-23, emphasis added)

Obviously, it does not make any sense to place death before original sin. As Christians, we need to know how to answer questions from unbelievers about death and suffering. It is only through repentance and faith in Christ that people can truly be comforted in the end.

Why do bad things happen to good people? (Biblical Answer)

There are 2 components to this question:

1. *There is no such thing as a good person, other than Jesus Christ, according to the Bible.* This book should have made that very clear!

"For all have sinned and fall short of the glory of God." (Romans 3:23)

"As it is written: 'None is righteous, no, not one; no one understands; no one seeks for God. All have turned aside; together they have become worthless; no one does good, not even one.'" (Romans 3:10-12)

"He (Jesus) committed no sin, neither was deceit found in his mouth." (1 Peter 2:22)

"You know that He (Jesus) appeared in order to take away sins, and in Him there is no sin." (1 John 3:5)

2. *Bad things happen in this world due to Adam and Eve, not God, as answered in the previous question* (Romans 8).

Why do bad things happen to good people? (Presuppositional answer)

Oftentimes, this question regarding bad things (death, disease, suffering, famine, heartache, etc.) will be asked in a variety of other ways, such as:

"Why do children get cancer?"
"Why do school shootings happen?"
"Why did my child die?"

You could answer the challenge by explaining Genesis 2 and Genesis 3 right at the beginning, but that will most likely be interpreted through that person's own worldview. The other way to answer the question would be to take the Presupp approach.

The person asking the question is invoking a sense of good and bad and/or right and wrong by asking the question. Instead, ask the person, *"You are invoking a standard of good and bad in your question. In your evolutionary worldview, where you are the result of random chemical reactions, what is your basis for an absolute standard of right and wrong?"* Now, you can continue on the logic here as covered in chapters 7 and 8.

What is your absolute standard for measuring what is good and what is evil (Right from Wrong)?

When a person is challenged on what his basis is for an absolute standard for morality, he will give you an answer that will fall under one or more of the following four options (as adapted from the book, *The Rights Fight*[4]):

1. Personal opinion
2. For the Benefit of society
3. Laws of society
4. Feelings

These 4 options all fall short - they are all relative standards. They have no universal, absolute standard! You can show this to the person by bringing up examples that contradict each one of these options.

1. Give him your opinion on something that is different than his. Then ask him who is right.
2. Ask him if what Hitler did is correct. His belief in the purification of the human race was supposed to be to the benefit of society.
3. Ask them if the slavery of black people was okay in the 1800's or the discrimination of races in the 1900's was okay, since the law permitted those things. In addition, how can laws change if they reflect an absolute standard?
4. Give him your feelings on something that is different than his regarding any stance that he takes. Then ask him who is right.

It is quite obvious that the only way to have an absolute standard of what is good and evil (right and wrong) comes from God, the absolute moral authority! Without the objective standard (God), the person only has preferences!

If God is so powerful and so good, then why do bad things happen in the world?

As answered earlier, bad things happen because of Adam's sin, not God.

If He is so powerful and so good, then why does God not stop all evil and bad things right now? The answer is simple- it is because of His grace and mercy. If He ended all evil right now, the world would end immediately and every person who is not saved would go to hell and suffer for eternity. It is because of His grace and mercy that He is still giving people time to repent and put their trust in Christ alone!

Atheist asks the question: "How do you know that God exists?"

Your response: The same way you do, but I am following him!" (Romans 1) Then start witnessing to him, as we outline in chapter 16.

Refuting post-modernism (relativism)

During encounters today, especially with the 30-somethings and younger, you will find that they are post-modern (relativist) in their mindset. You will usually hear some variation of one of the following retorts:

"Your truth is your truth, my truth is my truth."

"We can both be right."

"If that is what you believe."

Once I realize that the person I am encountering is post-modern, I will often point out:

"How can that be true? You can't have two opposing truths both be right!"

After a slight pause, I ask the question,

> "How do street lights change?"

(They will give you a logical answer regarding some type of timing device.)

Then, I say something wacky like this:

> "Well, I don't believe that. I believe that a little monkey lives in every streetlight and changes the colors. Now tell me that I am wrong."

The funny thing is, there are many people who will stick to the post-modern mindset and tell you one of the retorts above. Either way, you can point out the error in the logic! One of the interesting things to point out to the person is that he isn't actually a consistent relativist. After he responds in a post-modernist way, point out to him that he doesn't live his life accordingly. Asking a simple question can point out his inconsistent logic:

> "When you go to pay for lunch, and the cashier tells you that you owe $5.99, you don't give the cashier $2.00 and walk away."

If he is a college student, ask this question with the following answer:

> "When you have to pay your tuition at the beginning of the semester, do you just pay the cashier whatever you want? Of course you don't. You pay the amount that they tell you to. There certainly is absolute truth when it comes to what you owe for school!"

Even in the silly "monkey in the stoplight" example above, it can be used to demonstrate absolutes. The same person who would tell me that I can be right in my belief regarding the monkey changing stoplights would not agree with me if I said that red does not actually mean "stop," but it really meant to speed up.

Hopefully, the post-modernist will see the foolishness of his ways!

Refuting False Religions/Cults

Examples of false religions/cults include: Mormonism, Roman Catholicism, Islam, Jehovah's Witnesses, Buddhism, Hinduism, and many others. Remember that many borrow from God's word and then distort it. You can show how the false religion that he brings up violates the AIP test as explained in chapter 7.

Applying the AIP test to every religious worldview:

1. It cannot be arbitrary (without justification).
    a. The majority of religious systems have some element of being arbitrary, making it unsupported and false.
2. It cannot be inconsistent.
    a. Every religious system besides Christianity has contradictions within its doctrine, making it unreliable and false.
3. It must satisfy the preconditions of intelligibility (such as having a basis for absolute knowledge, and absolute morality)
    a. Every non-religious worldview lacks an absolute authority (God), thus none of them can satisfy this requirement.

Of all the religious systems and non-religious systems that exist, only Biblical Christianity can satisfy the necessary requirements to be a valid worldview. For more information on how to refute specific false religions, books by Pastor Andrew Rappaport[5] (strivingforeternity.org) and Bodie Hodge[6] are excellent resources.

Chapter Sixteen

Strategy of an Encounter - and How to Use This Knowledge

Encountering the Lost

Without turning evangelism into a recipe to follow, we want to provide some "architecture" to show how an evangelism encounter can be conducted simply to allow you to gain confidence. In the Great Commission, Jesus states:

> "Go therefore and make disciples of all nations, baptizing them in the name of the Father, the Son, and the Holy Spirit." (Matthew 28:19)

In understanding this verse, the "go" translates to "as you go." Everywhere we go, whether we are at work, at the mall, at the park, in church, etc., we are to be sharing the Gospel (evangelizing), teaching, making disciples, and baptizing. The easiest way to carry out the command of the Great Commission is by always carrying good Gospel tracts and handing them out to people. What makes up a good Gospel tract? It must contain a message that:

1. Uses the law to show a person his sins,
2. Shows that because he sins, he is a lawbreaker of God's law,
3. Teaches that God is Holy, Righteous, and Just- He must punish sin
4. We will all face Him on Judgment Day
5. We will go one of two places after that- Heaven or hell for eternity

6. We will only go to Heaven by repenting of our sin and putting our trust in Christ, who lived the law perfectly, and that He paid the penalty for our sins (He died on a cross for our sins, was buried, was raised 3 days later, conquering death - 1 Corinthians 15:3-4)

When you get bolder in your witness, you can start a conversation with someone- whether it is a family member, friend, or a random person. Sometimes this occurs while giving a tract to a person; sometimes I just walk up to someone and start a conversation. I use a variation of the same opening line every time I encounter someone, "How are you today? Did you ever think about what happens when you die?" Other people choose to use "spiritual transitions," as taught by Pastor Andrew Rappaport of Striving for Eternity Ministries (examples can be found on the ministry homepage: strivingforeternity.org). Someone employing this technique will use a recent news item or observation in a conversation with an unbeliever, and then transition that conversation to the Gospel.

Conversation started – Now what?

Once the conversation is started, we need to remember two things we know to be true from Scripture:

- Every person knows deep down that the Christian God exists by His creation, and that the unbeliever is only suppressing the truth about God because of his sin. He will be without excuse on Judgment Day. (Romans 1)
- God's Law is written on every person's heart. An absolute knowledge of good and evil is already present within that person. (Romans 2)

Apologetics goes hand-in-hand with Gospel proclamation- they are not two separate realms. We are to use apologetics to bring down strongholds, as we see in Paul's second letter to the Corinthians:
"For though we walk in the flesh, we are not waging war according to the flesh. For the weapons of our warfare are not of the flesh but have divine power to destroy strongholds. We

destroy arguments and every lofty opinion raised against the knowledge of God, and take every thought captive to obey Christ." (2 Corinthians 10:3-5)

Understand, though, that *the goal of every encounter is to use apologetics minimally, while getting to the Law and Gospel as quickly as possible.* It is our job to be obedient in speaking God's Word faithfully and accurately. However, we are not responsible, nor even have the power, to open up or change an unbelieving heart. Only the Holy Spirit can change hearts. It is key to understand that no one is saved by apologetics- people are only saved by the changing power of the Gospel. As we read in Ezekiel 36:

"And I will give you a new heart, and a new spirit I will put within you. And I will remove the heart of stone from your flesh and give you a heart of flesh." (Ezekiel 36:26)

The basic outline of an encounter should resemble something like this:

1. Remember what we know about everyone when walking into an encounter: He knows the true God that exists (Romans 1) and His law is written on the person's heart (Romans 2).
2. Ask questions to determine the person's worldview (evolution or false religion).
3. Use presuppositional apologetics and AIP test to refute the person's non-Christian worldview.
4. Use some evidential apologetics only if the person has 1 or 2 sincere questions. Do not get caught in a debate- this is typically just a time waster.
5. Preach the Law
6. Proclaim the Gospel

Remember- do not fall in love with the apologetic method. You will always win the argument, but it does not save anyone... You need to get to the Law and preach the Gospel!

Sample Encounter (Great examples can be found on Livingwaters.com)[1]

1. Pray for any and all encounters that the Lord opens the door for

2. Remember what we know about everyone when walking into an encounter: He knows the true God that exists and His law is written on the person's heart.

3. Lock eyes with the person you are encountering to show you care.

4. Say:
    a. "Hello!"
    b. "How are you doing today?"
    c. "Have you gotten one of these yet?" (Hand the person a Gospel tract)
    d. "I have a question for you if you have a minute. Have you ever thought about what happens when you die?"

5. Usually the person will say something in the affirmative.
    a. If so, ask, "Where do you think you will go, Heaven or hell?"
    b. If the person does not answer in the affirmative, then say, "Where do you think you will go, Heaven or hell?"
    c. Notice that you can still transition to this question no matter how the person answers!

6. The person will answer either: Heaven, hell, or neither of those options.
    a. If Heaven, ask, "Why, Heaven?" (Go to #8)
    b. If hell, ask, "Why hell? (Go to #7)

    c. If the answer is something else (reincarnation, we become dirt, nothing, ...) then say, "If you are pulled over for speeding, it will do no good to tell the officer that you do not believe in speed limits. So, will you go to Heaven or hell?"

7. If the person answered hell, it will almost always be because the person understands that he is not a good person, one way or another.
    a. Say, "Does that concern you? Would you like to know how you can be saved and go to Heaven?" (Go to #9 to make sure the person understands the definition of good by God's standards)

8. If the person answered Heaven, it will almost always be because the person either says to you outright that he is a good person, or he will say a lot of things that will make the case that he is a good person, without stating it outright.
    a. Ask the person if he would like to take "The Good Person Test" with you. They almost always say yes.

9. Give "The Good Person Test" (adapted from Ray Comfort, Living Waters Ministry) Ask,
    a. "Have you ever told a lie in your life? I know that I have!"
        i. After they answer, say, "What do you call someone who has ever told a lie?" Get him to say, "A Liar"
    b. Have you ever stolen anything in your life? I have also done this!"
        i. After they answer, say, "What do you call someone who has ever stolen anything?" Get him to say, "A Thief"
    c. Have you ever hated someone before, as Jesus calls hate committing murder in your heart?"
        i. After they answer, say, "I am guilty of this."
    d. Have you ever lusted, had impure sexual thoughts, after anyone before? Jesus calls that committing adultery in your heart."

i. After they answer, say, "I am also guilty of this."

10. Say, "So you and I are each liars, thieves, murderers, and adulterers at heart. Will you be found innocent or guilty? Where do you think that you will end up on Judgment Day?
    a. If he says either "innocent" or "heaven," (Go to #11)
    b. If he says either "guilty" or "hell," ask him if he would like to know how you can be saved and go to Heaven." (Go to #12)

11. Say, "Why do you still think you will go to Heaven?" He will usually answer that God is forgiving, that he is still not that bad, or that the bad outweighs the good. Regardless of the answer, give him the following scenario.
    a. Say, "Imagine that you are 30 years old. (Always guess an age that is younger than what you think he is.) Let's say that you lived a perfect life- you never sinned or broke a law- until you were 25. Then, 5 years ago, you robbed a bank, got away with a million dollars, and never got caught. Then you go 5 more years with not sinning or breaking another law. Thus, you have only done one thing wrong your entire life – robbing that bank. Now, imagine that some new evidence comes out and you get arrested. You are now facing the judge, a *fair and just judge*. He will ask, "We know that you did this. How do you plead to these charges?" Obviously you will plead guilty. Will you be able to get out of the punishment be telling the judge that you are a good person? Will you be able to say, "Judge- I hold open doors for people. I walk old ladies across the street. I donate time. I donate lots of money. I treat people nice. Can you let me off?" If the judge is fair and just, will he be able to let you off without punishment? Of course not! How much more do you think that the God of the universe, who created you and I, must punish us for breaking His law? (Go to #12)

12. Give the Gospel
    a. Say, "Every single person, including you and I, deserve an eternity in Hell for our law breaking, The good news of the Gospel is that God, who must punish all lawbreakers, sent His Son Jesus Christ to die on a Cross for us and pay the debt that we owe for our sin. Jesus Christ was perfect- He lived the law perfectly on earth, something that you and I are incapable of. He was beaten, tortured, and died on a cross. He was buried and then raised three days later, conquering death. You must repent of your sin and put your trust in Christ to be saved from hell!"

Answering Challenges

During the conversation, the person may try to change the direction. It is of the utmost importance to keep command of the conversation, stay on track, and go through the Law and give the Gospel! When someone asks a question, always remember to stick to a presuppositional framework to respond. It is ok to answer a few *sincere* questions that he may have about Christianity, but make sure he is not trying to just waste your time. Then, get back to the Law and the Gospel.

Answering Worldview Challenges

When questioned about different worldviews, remember the AIP test. Every religion besides Christianity will fail the "A", the "I", or both. Every non-religious worldview has evolution at its core. It will always fail the "P", and most fail the "A" and "I" as well.

The majority of people today have a post-modern mindset. Truth is relative in their minds. This is also easy to expose just by asking questions and pointing out the logic problems.

Questions to Ask an Evolutionist (Adapted from Mike Riddle, of CTI)

Memorize these 4 power questions to ask an evolutionist to break down his worldview:

1. Where did the matter come from that created the big bang?
2. How did life start?
3. What about the Fossil Record? Where are the millions of transitions between the Cambrian and Precambrian periods?
4. Where did the dinosaurs come from?

These questions are not answerable by the evolutionist. Here is why:

1. We addressed this in chapter 13. They cannot account for it, yet they criticize creationists for understanding that God has always existed and transcends time.

2. They have no explanation for this, and the Miller experiment is faulty, as explained in chapter 13.

3. They have labeled this the Cambrian Explosion. Again, this is a bogus explanation as to why there are millions of missing transitions, and was answered in chapter 13.

4. They can't account for where dinosaurs came from.

Questions to ask an Old-Earth Professing Christian

1. Where in the Bible is millions/billions of years mentioned or insinuated?

2. Where in the Bible can you find that God used evolution to create Adam and Eve?

3. Where in the Bible can you find God changing one of His created kinds into another kind of animal?

4. Where did thorns and thistles come from?

These questions are not answerable by the Old-Earth professing Christian. Here is why:

1. This isn't found anywhere in the Bible. As explained earlier, Old Earth professing Christians have to jam this arbitrarily into the Bible. Their only source for MOY (Millions of Years)/ BOY (Billions of years) is from some scientists.

2. This isn't found anywhere. God said that he made Adam from the dirt and Eve from Adam's rib. And both were created in the image of God – not a monkey!

3. This is also not present anywhere in the Bible as God continually said that he made things to reproduce after its own kind. This was explained in chapter 3.

4. Thorns and thistles appear throughout the fossil record, which old-earth Christians believe to be millions-to-billions of years old. Yet, the Bible is clear that thorns and thistles only appeared *after* Adam's sin in the Garden, about 6000 years ago, according to Genesis 3:18. This is yet another proof that the fossil record confirms a young earth!

Conclusion: Refuting anything that anyone says

The 2-Move Checkmate (Adapted from Sye Ten Bruggencate) is what we retort when someone makes a truth claim or knowledge claim that does not line up with the Scripture:

> "That is not what the Bible says. How do you get truth without God?"

We can follow that up with the logic that we learned in Chapter 7. Because **God *is* truth**, we must compare everything that we hear to His Word. When something that we hear does not match up with the Word of God, we are called to challenge it. As we have established throughout this book, the Christian worldview is the only correct one. We can quote Scripture at anytime and know that we are preaching **truth.**

> "For the word of God is living and active, sharper than any two-edged sword, piercing to the division of soul and of spirit, of joints and of marrow, and discerning the thoughts and intentions of the heart." (Hebrews 4:12)

Always call people back to the Word of God, the maker of Heaven and Earth. Jesus is the Word, the Truth, and the Creator of everything, including the person you are witnessing to!

We pray that this book has been a blessing to you!

"Quick Quips"
Quick responses to the unbeliever

(These are not meant to insult; they are meant only to close the mouth of the heckler and share the Gospel)

Unbeliever: If your God is real, then why doesn't he provide evidence and show himself to us?
You: He did that 2000 years ago, and they nailed Him to a cross!
(courtesy of Ken Ham, AIG)

When being heckled...
You: Do you go to the mall and heckle Santa Claus while all the little kids are in line?
Unbeliever: No.
You: Why not?
Unbeliever: Because he isn't real.
You: Exactly. Yet you are heckling me because you know that God is real!
(courtesy of Sye Ten Bruggencate, ProofThatGodExists.org)

Unbeliever: I don't believe in absolutes (absolute truth, absolute knowledge)
You: Are you absolutely sure?

Unbeliever: I don't believe in absolutes (absolute truth, absolute knowledge)
You: How do you even believe the accuracy of the thoughts and words coming out of your mouth?

Unbeliever: I don't believe in absolutes (absolute truth, absolute knowledge)
You: Yet, you want us to believe in what you have been saying?

Unbeliever: How do you know that God exists?
You: The same way that you do, but I am following Him!

Professing Christian: I believe there may be some errors in the Bible.
You: Then when did God start telling us the truth?

Unbeliever: The Bible is a fairy tale!
You: What do you call a story that begins with "Long ago and far away?"
Unbeliever: A fairy Tale!
You: Correct! What do you call a story that begins with "A long time ago?"
Unbeliever: A fairy Tale!
You: Correct! What do you call a story that begins with "Once upon a time?"
Unbeliever: A fairy Tale!
You: The Bible starts with, "In the beginning, God." Your story begins with, "13.5 billion years ago!" Now which one sounds like a fairy tale to you?
(Courtesy of Dr. Anthony Silvestro, Creation Revival)

Recommended Resources to Supplement What is Learned While Reading this Book (In no particular order)

Creation Resources:

Creation Training Initiative (CTI), Mike Riddle
Answers In Genesis (AiG)
Creation Ministries International (CMI)
Institute for Creation Research (ICR)
Creation Today, Eric Hovind
John MacArthur

Presuppositional Apologetics Resources:

Dr. Jason Lisle, *The Ultimate Proof of Creation*
Sye Ten Bruggencate, *How to Answer the Fool* (video)
Jeff Durbin (Apologia Radio and TV)
Todd Friel (Wretched Radio and TV)
Mark Spence (Living Waters Ministry)
Greg Bahnsen, *Always Ready*
Greg Bahnsen, *The Great Debate: Does God Exist?*
Jay Lucas, *Ask Them Why*
Jay Lucas, *The Rights Fight*
CARM.org, Christian Apologetics Research Ministry, Reverand Matt Slick

Biblical Hermeneutics

Striving for Eternity Ministries -Pastor Andrew Rappaport

Systematic Theology

Striving for Eternity Ministries -Pastor Andrew Rappaport

Footnotes (By Chapter)

Chapter 1: The Christian Worldview (Dr. Anthony R. Silvestro, Jr.)

1. http://www.ligonier.org/learn/devotionals/inspiration-infallibility-inerrancy/
2. www.apologeticspress.org, *Is the New Testament "Given by Inspiration of God"?*, by Eric Lyons M.Min.
3. https://carm.org/bible-inspired
4. http://www.biblestudytools.com/dictionaries/bakers-evangelical-dictionary/sanctification.html
5. http://www.ligonier.org/blog/what-does-it-mean-fear-god/

Chapter 2: Origins (Dr. Anthony R. Silvestro, Jr.)
1. http://creation.com/the-date-of-noahs-flood
2. http://www.penn.museum/games/cuneiform.shtml
3. http://humanorigins.si.edu/education/intro-human-evolution
4. http://www.britannica.com/science/taxonomy/The-Linnaean-system

Chapter 3: Basic Evolutionary "Science" (Dr. Anthony R. Silvestro, Jr.)

1. http://www.merriam-webster.com/dictionary/science
2. http://sciencecouncil.org/about-us/our-definition-of-science/
3. http://www.bju.edu/academics/college-and-schools/arts-and-science/natural-science/teaching-science/definition.php
4. http://creation.com/origins-vs-operational-science
5. https://answersingenesis.org/what-is-science/what-is-science/
6. https://answersingenesis.org/what-is-science/two-kinds-of-science/
7. http://sciencecouncil.org/about-us/our-definition-of-science/

8. http://physics.ucr.edu/~wudka/Physics7/Notes_www/node6.html
9. http://creation.com/its-not-science
10. https://answersingenesis.org/environmental-science/climate-change/a-proposed-bible-science-perspective-on-global-warming/
11. http://hubblesite.org/reference_desk/faq/all.php.cat=cosmology
12. http://www.scientificamerican.com/article/how-science-figured-out-the-age-of-the-earth/
13. http://www.talkorigins.org/faqs/geohist.html
14. http://creation.com/dinosaur-soft-tissue
15. http://ncse.com/rncse/21/1-2/defining-evolution
16. http://biology.about.com/od/evolution/a/aa110207a.htm
17. http://science.nationalgeographic.com/science/prehistoric-world/mass-extinction/
18. http://www.britannica.com/science/taxonomy
19. http://www.txstate.edu/philosophy/resources/fallacy-definitions/Equivocation.html
20. http://www.medicinenet.com/genetic_disease/article.htm
21. http://www.netwellness.org/healthtopics/idbd/2.cfm

Chapter 4: The Relevance of Genesis (Dr. Anthony R. Silvestro, Jr.)

1. www.creationstudies.org/Education/quotations.html
2. *Revolutionary Parenting*, George Barna
3. *Already Gone,* Britt Beemer, Ken Ham
4. http://www.gty.org/resources/study-guides/40-5129/love-not-the-world
5. http://www.nizkor.org/hweb/imt/nca/nca-01/nca-01-07-means-46.htm
6. *The Communist Manifesto*, Karl Marx, point 18
7. https://nces.ed.gov/surveys/sass/tables/sass0708_035_s1s.asp
8. *http://www.bls.gov/TUS/CHARTS/LEISURE.HTM*
9. *Jesus Unmasked*, Todd Friel, pgs. 115-116
10. https://answersingenesis.org/bible-questions/why-do-you-take-the-bible-literally/

Chapter 5: Creation, the Fall, and the Promise (Dr. Anthony R. Silvestro, Jr.)

Chapter 6: The Gospel (Jonathan Eckel)

1. *The Bible*, God
2. http://av1611.com/kjbp/kjv-dictionary/imputable.html

Chapter 7: An Introduction to Presuppositional Apologetics (Dr. Anthony R. Silvestro, Jr.)

1. *The Ultimate Proof of Creation*, Dr. Jason Lisle
2. http://www.merriam-webster.com/dictionary/arbitrary
3. https://school.carm.org/amember/files/demo3/2_logic/3logic.htm
4. http://study.com/academy/lesson/the-three-laws-of-logic.html

Chapter 8: Problems with Evolution (Dr. Anthony R. Silvestro, Jr.)

1. *A Universe From Nothing: Why There is Something Rather Than Nothing*, Lawrence M. Krauss
2. *The Rights Fight*, Jay Lucas

Chapter 9: The Reliability of the Bible (Dr. Anthony R. Silvestro, Jr.)

1. https://www.youtube.com/watch?v=G1XJ7DeR5fc, Why I Choose to Believe the Bible, Voddie Baucham
2. https://www.icr.org/article/480/
3. *More Than a Carpenter*, Josh McDowell
4. https://bible.org/seriespage/4-bible-written-word-god
5. http://www.biblicalarchaeology.org/category/daily/biblical-artifacts/dead-sea-scrolls/

Chapter 10: Biblical Reliability of the Text (Pastor Andrew Rappaport)

    1. http://danielbwallace.com/2013/09/09/the-number-of-textual-variants-an-evangelical-miscalculation
    2. http://danielbwallace.com/2013/09/09/the-number-of-textual-variants-an-evangelical-miscalculation
    3. Aslan, Reza, "Zealot: The Life and Times of Jesus of Nazareth" (Random House, 2013), XXVII.
    4. https://carm.org/bible-text-manuscript-tree
    5. Wallace, D. B. (2011). Lost in Transmission: How Badly Did the Scribes Corrupt the New Testament Text? In D. B. Wallace (Ed.), *Revisiting the Corruption of the New Testament: Manuscript, Patristic, and Apocryphal Evidence* (p. 28). Grand Rapids, MI: Kregel Academic & Professional.

Chapter 11: How the Gospel is Affected by the Evolutionary Argument (John Eckel)

    1. http://www.iep.utm.edu/evolutio/

Chapter 12: A Call to Repentance (Jonathan Eckel)

    1. *Ephesians, Volume 2,* by Dr. Martyn Lloyd-Jones, p.14

Chapter 13: Basic Challenges – Age of the Earth (Dr. Anthony R. Silvestro, Jr.)

    1. http://creation.mobi/biblical-age-of-the-earth, Lita Cosner
    2. Creation Training Initiative, Teachers' Training Course, Mike Riddle (Research cited from CMI, AIG, and ICR)

Chapter 14: Basic Challenges – Part 1 (Dr. Anthony R. Silvestro, Jr.)

    1. http://www.icr.org/article/statistical-determination-genre-biblical/

2. *On the Origin of Species by Means of Natural Selection, or the Preservation of Favoured Races in the Struggle for Life,* Charles Darwin, Chapter X
3. Creation Training Initiative, Teachers' Training Course, Mike Riddle (Research cited from CMI, AIG, and ICR)
4. *Noah's Ark: A Feasibility Study,* John Woodmorappe,
5. https://answersingenesis.org/genetics/fresh-look-human-chimp-dna-similarity/
6. http://creation.com/why-the-miller-urey-research-argues-against-abiogenesis
7. https://answersingenesis.org/theory-of-evolution/evolution-timeline/cambrian explosion- was-the-culmination-of-cascading-causes-evolutionists-claim/
8. https://answersingenesis.org/geology/radiometric-dating/radiometric-dating-back-to-basics/
9. https://answersingenesis.org/theory-of-evolution/millions-of-years/rate-overturning-millions-years/
10. http://www.icr.org/article/a-30-years-later-lessons-mount-st-helens/

Chapter 15: Basic Challenges – Part 2 (Dr. Anthony R. Silvestro, Jr.)

1. https://answersingenesis.org/tower-of-babel/was-the-dispersion-at-babel-a-real-event/
2. www.creation.com/noah-and-genetics
3. http://creation.com/light-travel-time-a-problem-for-the-big-bang
4. *The Rights Fight*, Jay Lucas
5. *What Do They Believe,* Andrew Rappaport
6. *World Religions and Cults, Volumes 1&2,* Bodie Hodge and Roger Patterson

Chapter 16: Strategy of an Encounter - and How to Use This Knowledge (Dr. Anthony R. Silvestro, Jr.)

1. Adapted from Living Waters Evangelism Training materials, including the Basic Training Course, Ray Comfort

About the Authors

Dr. Silvestro was once an unsaved evolutionist but now is an avid Biblical creationist – teaching adults and kids alike that God didn't create man from baboons, and that the Bible can be trusted from beginning to end, starting with the book of Genesis!

As a graduate of Mike Riddle's Creation College (former speaker for Answers in Genesis), and after years of study, Dr. Silvestro speaks on Biblical Creation and how science supports the Biblical account of creation as well as the age of the earth. Some of these topics include: The Relevance of Genesis, Noah's Ark and Flood, The Lie of Evolution, The Significance of the Tower of Babel, dinosaurs, and radiometric dating methods. Christians have benefitted by his teaching at conferences, youth groups, and Christian radio shows. He will also come to your church to train your church members to teach others through an abbreviated 1 day class or an extended 12-hour training class, whether it is over a two-day weekend or a 12-week period!

Dr. Silvestro also enjoys teaching presuppositional apologetics, the biblical apologetic method, to help equip Christians to proclaim the Gospel.

Dr. Anthony R Silvestro, Jr., resides in the Cleveland, OH area with his wife Julie, and his son, Anthony III. After receiving his Bachelor's Degree in Math and Chemistry from Baldwin-Wallace College, he attended The Ohio State University College of Dentistry and received his dental degree. Dr. Silvestro practices general dentistry and is also trained in most of the dental specialties.

Dr. Silvestro can be contacted by email or phone to book a training seminar at your church!

Email: creationrevival@gmail.com

About the Authors

John is a husband to Elena and a father to five beautiful children. John is passionate to live radically for Christ and to share the true Gospel with all he comes in contact with.

In 2013, the Lord put it on his heart to start a ministry called Remember Ministries. The purpose of the ministry is to reach out to those on the streets of every city he can, to love them, and to share God's supernatural Gospel that can transform empty lives into vessels used by Christ, for His Glory. Matthew 28:19 says, "Go therefore and make disciples of all nations, baptizing them in the name of the Father and of the Son and of the Holy Spirit."

www.rememberministries.org

Dr. Anthony Silvestro and Jonathan Eckel have written a book that is theologically grounded, biblically faithful, and evangelistically driven. This volume will answer questions like: Why does the only credible worldview begin with God and His self-revelation in Scripture? What does the Garden of Eden have to do with a place called Golgotha? How can I contend for the faith and give an answer to those who are opposed to the truth of the gospel?

Too many Christians do not posses the conviction to preach the gospel to the lost. Nor do they feel equipped to answer all of the objections to God's truth. We cannot assume that the people we will meet on a daily basis have any understanding of God or the Bible. This book will instruct the reader with a gospel-centered, Bible-saturated worldview and show any Christian how they can earnestly content for the faith, starting with the book of Genesis. The Christian will be encouraged with a helpful explanation of the "science" behind the non- Christian worldview. Questions like the age of the earth, the origin of life, and the problems with evolution, are carefully explained. Chapter 6 is particularly helpful in an age where gospel confusion abounds in the church. After reading this book, the Christian will have a better understanding of what he believes, how he can defend the faith, and how he can engage anyone in a gospel conversation. I heartily endorse this work and pray it will be used by God to edify and equip the saints in the work of the great commission.

For the glory of Christ and His church

Pastor Chris Hinckley
Olmsted Falls Baptist Church